Physically Spea

A Dictionary of Quot
on Physics and Astro1.___.y

About the Compilers

Carl C Gaither was born in 1944 in San Antonio, Texas. He has conducted research work for the Texas Department of Corrections and for the Louisiana Department of Corrections. Additionally he has worked as an Operations Research Analyst for the past ten years. He received his undergraduate degree (Psychology) from the University of Hawaii and has graduate degrees from McNeese State University (Psychology), North East Louisiana University (Criminal Justice), and the University of Southwestern Louisiana (Mathematical Statistics).

Alma E Cavazos-Gaither was born in 1955 in San Juan, Texas. She has worked in quality control, material control, and as a bilingual data collector. She received her associate degree (Telecommunications) from Central Texas College.

They arranged and compiled the book *Statistically Speaking*.

About the Illustrator

Andrew Slocombe was born in Bristol in 1955. He spent four years of his life at Art College where he attained his Honours Degree (Graphic Design). Since then he has tried to see the funny side to everything and considers that seeing the funny side to Physics has tested him to the full! He would like to thank Carl and Alma for the challenge!

Physically Speaking
A Dictionary of Quotations on Physics and Astronomy

Selected and Arranged by

Carl C Gaither
and
Alma E Cavazos-Gaither

Illustrated by Andrew Slocombe

Published in 1997 by
Taylor & Francis Group
270 Madison Avenue
New York, NY 10016

Published in Great Britain by
Taylor & Francis Group
2 Park Square
Milton Park, Abingdon
Oxon OX14 4RN

International Standard Book Number-10: 0-7503-0470-7 (Softcover)
International Standard Book Number-13: 978-0-7503-0407-2 (Softcover)

This book contains information obtained from authentic and highly regarded sources. Reprinted material is quoted with permission, and sources are indicated. A wide variety of references are listed. Reasonable efforts have been made to publish reliable data and information, but the author and the publisher cannot assume responsibility for the validity of all materials or for the consequences of their use.

Library of Congress Cataloging-in-Publication Data

Catalog record is available from the Library of Congress

Taylor & Francis Group
is the Academic Division of Informa plc.

Visit the Taylor & Francis Web site at
http://www.taylorandfrancis.com

This book is dedicated to the ones I love,
my sisters and brothers
Sonny, Tito, Witcha, Rosie, Mike, Corny,
Jelly, Negro, Marta, and Margot

Alma E Cavazos-Gaither

This book is dedicated to my friend
David Adams Puzzuoli

Carl C Gaither

Let playful PENDULES quick vibration feel,
While silent CYCLOIS rests upon her wheel;
Let HYDROSTATICS, simpering as they go,
  Lead the light Naiads on fantastic toe;
  Let shrill ACOUSTICS tune tiny lyre;
With EUCLID sage fair ALGEBRA conspire;
The obedient pulley strong MECHANICS ply,
  And wanton OPTICS roll the melting eye!

Frere, C.
Canning, B.
In Charles Edmund's
Poetry of the Anti-Jacobian
The Loves of the Triangle
Canto I, l. 13–20

CONTENTS

PREFACE

Physically Speaking: A Dictionary of Quotations on Physics and Astronomy has, for the first time, brought together in one easily accessible form the best expressed thoughts that are especially illuminating and pertinent to the disciplines of physics and astronomy. Some of the quotations are profound, others are wise, some are witty, but none are frivolous. Quotations from the most famous men and women lie in good company with those from unknown wits. You may not find all the quoted 'jewels' that exist, but we are certain that you will find a great number of them here. We believe that Benjamin Franklin was correct when he said that "Nothing gives an author so much pleasure as to find his work respectfully quoted...".

Frequently, books on quotations will have subtle changes to the quotation, changes to punctuation, slight changes to the wording, even misleading information in the attribution, so that the compiler will know if someone used a quotation from 'their' book. We attempted to verify each and every one of the quotations in this book to ensure that they are correct.

The attributions give the fullest possible information that we could find to help you pinpoint the quotation in its appropriate context or discover more quotations in the original source. Speeches include, when possible, the date of the speech. We assure the reader that not one of the quotations in this book was created by us.

In summary, *Physically Speaking* is a book that has many uses. You can:

- Identify the author of a quotation.
- Identify the source of the quotation.
- Check the precise wording of a quotation.
- Discover what an individual has said on a subject.
- Find sayings by other individuals on the same subject.

How to Use This Book
1. A quotation for a given subject may be found by looking for that subject in the alphabetical arrangement of the book itself. To illustrate,

if a quotation on "electrons" is wanted, you will find twenty-six quotations listed under the heading electron. The arrangement of quotations in this book under each subject heading constitutes a collective composition that incorporates the sayings of a range of people.

2. To find all the quotations pertaining to a subject and the individuals quoted use the SUBJECT BY AUTHOR INDEX. This index will help guide you to the specific statement that is sought. A brief extract of each quotation is included in this index.

3. If you recall the name appearing in the attribution or if you wish to read all of an individual author's contributions that are included in this book then you will want to use the AUTHOR BY SUBJECT INDEX. Here the authors are listed alphabetically along with their quotations. The birth and death dates are provided for the authors whenever we could determine them.

Thanks

It is never superfluous to say thanks where thanks are due. First, I again wish to thank my stepdaughter Maritza Marie Cavazos for her assistance in tracking down incomplete citations, looking for books in the libraries, and helping to sort the piles of correspondence generated in obtaining permissions. I thought that she would have been burnt out after working so hard with us on *Statistically Speaking* but she wasn't. Additionally, Alma and I wish to thank Ammah Cabrera for her help with the library work. It was really appreciated. Next, we thank the following libraries for allowing us to use their resources: The Jesse H. Jones Library and the Moody Memorial Library of Baylor University; the Townsend Memorial Library of the University of Mary Hardin-Baylor; the Oveta Culp Hobby Memorial Library of the Central Texas College; the Perry-Castañeda Library, the Undergraduate Library, the Engineering Library, the Law Library, the Physics-Math-Astronomy Library, and the Humanities Research Center of the University of Texas at Austin. Finally we wish to thank our publisher Jim Revill for his assistance in the publication of this book.

Additionally, we would like to thank each of the publishers who provided permission to use the quotations. We made a very serious attempt to contact the publishers for permission to use the quotations. Letters were written to each publisher or agent for which we could find an address. If no response was received we then assumed a calculated risk and incorporated the quotation. In no way did we use a quotation without attempting to obtain prior approval.

Carl Gaither
Alma Cavazos-Gaither
July 1997

ANTI-MATTER

Furth, Harold P.
Well beyond the tropostrata
There is a region stark and stellar
Where, on a streak of anti-matter
Lived Dr. Edward Anti-Teller.

The New Yorker
The Talk of the Town
Perils of Modern Living (p. 56)
November 10, 1956

Heisenberg, Werner
I think that this discovery of antimatter was perhaps the biggest jump
of all the big jumps in physics in our century.

Quoted by J. Mehra (Editor) in
The Physicist's Conception of Nature (p. 271)

Schuster, Arthur
Astronomy, the oldest and most juvenile of the sciences, may still have
some surprises in store. May anti-matter be commended to its care!

Nature
Letter to the Editor
Potential Matter—A Holiday Dream (p. 367)
Volume 58, August 18, 1898

Updike, John
Think binary. When matter meets antimatter, both vanish, into pure
energy. But both existed; I mean, there was a condition we'll call
"existence." Think of one and minus one. Together they add up to zero,
nothing, nada, niente, right? Picture them together, then picture them
separating—peeling apart . . .. Now you have something, you have two
somethings, where you once had nothing.

Roger's Version (p. 304)

ASTRONOMER

Boyle, Robert
Arithmetic and geometry, those wings on which the astronomer soars as high as heaven.

Works
Usefulness of Mathematics to Natural Philosophy
Volume 3 (p. 429)

Browne, Sir James Crichton
We follow with awe and thrilling interest the prodigious revelation of our astronomers, but after all our conception of the Stellar Universe does not go much further than:

Twinkle, twinkle little star . . .

From the Doctor's Notebook
Matrimonial Obedience (p. 152)

The astronomers are wonderful and impressive in their own sphere, but when they stray into literature they become annoying. We have all rejoiced in that beautiful elegy on the Burial of Sir John Moore, and know the essential line:

By the struggling moonbeam's misty light.

Well, the astronomers have worked it out and found there was no moon the night of the Battle of Corunna.

From the Doctor's Notebook
The Astronomers (p. 192)

The astronomers with all their hypotheses give us no satisfying or abiding conception of the Universe. We are left as bewildered as ever.

From the Doctor's Notebook
Tea (p. 224)

D'Avenant, Sir William
You that so wisely studious are
To measure and to trace each Starr,
How swift they travaile, and how farr,
   Now number your celestiall store,
Planets, or lesser lights, and trie
If in the face of all the skie
   You count so many as before!

Salmacida Spolia
IIII. Song; Stanza I

de Fontenelle, Bernard
... as princes monopolize the earth, it is but fair that astronomers should have the sky for their share, and not suffer princes to intrude on their domain.

Conversations on the Plurality of Worlds
Fourth Evening (p. 89)

de Morgan, Augustus
Astronomers! What can avail
   Those who calumniate us;
Experiment can never fail
   With such an apparatus ...

Budget of Paradoxes
Volume I
The Astronomer's Drinking Song (p. 380)

de Saint-Exupéry, Antoine
This asteroid has only once been seen through the telescope. That was by a Turkish astronomer, in 1909.

On making his discovery, the astronomer had presented it to the International Astronomical Congress, in a great demonstration. But he was in Turkish costume, and so nobody would believe what he said.

The Little Prince
Chapter 4

D'Israeli, Isaac
It does at first appear that an astronomer rapt in abstraction, while he gazes on a star, must feel more exquisite delight than a farmer who is conducting his team.

Literary Character of Men of Genius
On Habituating Ourselves to an Individual Pursuit

Dryden, John
As the true height and bigness of a star
Exceeds the measures of th' astronomer.

The Poetical Works of Dryden
Eleonora
l. 264–5

The sun was enter'd into Capricorn;
Which, by their bad astronomer's account,
That week the Virgin Balance should remount . . .

The Poetical Works of Dryden
The Hind and the Panther
l. 1892

Dunne, Finley Peter
. . . I know about marredge th' way an asthronomer knows about th'
stars.

Mr. Dooley At His Best
Marriage

Emerson, Ralph Waldo
The astronomer, the geometer, rely on their irrefragable analysis, and
disdain the results of observation.

The Works of Ralph Waldo Emerson
Volume I
Nature
Idealism (p. 56)

Esar, Evan
[Astronomer] An anagram for *moon starer*.

Esar's Comic Dictionary

[Astronomer] A man whose business is always looking up.

Esar's Comic Dictionary

[Astronomer] The only night watchman who doesn't sleep on the job.

Esar's Comic Dictionary

[Astronomer] A man who looks at the moon even when he is not in love.

Esar's Comic Dictionary

[Astronomer] A fortunate man whose job is just to stand around and
look at heavenly bodies.

Esar's Comic Dictionary

Fort, Charles

. . . if nobody looks up, or checks up, what the astronomers tell us, they are free to tell us anything that they want to tell us.

The Books of Charles Fort
Lo
Part 2, Chapter XX

Frost, Robert

As a confirmed astronomer
I'm always for a better sky.

The Poetry of Robert Frost
A-Wishing Well
l. 12–13

Harington, Sir John

Astronomers, Painters and Poets may lye by authoritie.

Quoted by G. Gregory Smith in
Elizabethan Critical Essays
Sir John Harington
A Brief Apology for Poetry (p. 201)

Hodgson, Ralph

Reason has moons, but moons not hers
  Lie mirrored on her sea,
Confounding her astronomers,
  But O! delighting me.

Collected Poems
Reason has Moons

Howard, Neale E.

Astronomers work always with the past; because light takes time to move from one place to another, they see things as they were, not as they are.

The Telescope Handbook and Star Atlas
The Sky, Chapter III (p. 33)

Hoyle, Fred

The astronomer is severely handicapped as compared with other scientists. He is forced into a comparatively passive role. He cannot invent his own experiments as the physicist, the chemist or the biologist can. He cannot travel about the Universe examining the items that interest him. He cannot, for example, skin a star like an onion and see how it works inside.

The Nature of the Universe
The Earth and Near-By Space (pp. 3–4)

Hunstead, Richard
Astronomers like to plot things on logarithmic scales because the errors are so huge and the trends so weak that you have to plot them on a log–log graph in order to get anyone to believe them.

University of Sydney

Jeffers, Robinson
Nor can the astronomer see his moon-dazzled
Constellations: let him give one night in the month to
    earth and the moon,
Women and games.

The Beginning and the End
Full Moon

Kirshner, Robert
Like the fifteenth-century navigators, astronomers today are embarked on voyages of exploration, charting unknown regions. The aim of this adventure is to bring back not gold or spices or silks but something more valuable: a map of the universe that will tell of its origin, its texture, and its fate.

Quoted by Marcia Bartusiak in
Thursday's Universe (p. 167)

Mencken, H.L.
Astronomers and physicists, dealing habitually with objects and quantities far beyond the reach of the senses, even with the aid of the most powerful aids that ingenuity has been able to devise, tend almost inevitably to fall into the ways of thinking of men dealing with objects and quantities that do not exist at all, e.g., theologians and metaphysicians.

Minority Report: H.L. Mencken's Notebook
Sample 74

Prior, Matthew
At night astronomers agree . . .

The Literary Works of Matthew Prior
Volume VI
Phillis's Age
Stanza 3, l. 2

Prudhomme, Sully
Tis late; the astronomer in his lonely height
Exploring all the dark, descries from far
Orbs that like distant isles of splendor are . . .

Quoted by Morris Kline in
Mathematics in Western Culture (p. 60)

Rheticus, Georg Joachin
The astronomer who studies the motion of the stars is surely like a blind man who, with only a staff [mathematics] to guide him, must make a great, endless, hazardous journey that winds through innumerable desolate places. What will be the result? Proceeding anxiously for a while and groping his way with his staff, he will at some time, leaning upon it, cry out in despair to Heaven, Earth and all the Gods to aid him in his misery.

Quoted by Arthur Koestler in
The Sleepwalkers
Part III, I, 10 (p. 161)

Robinson, Edwin Arlington
And thus we die,
Still searching, like poor old astronomers,
Who totter off to bed and go to sleep
To dream of untriangulated stars.

Collected Poems of Edwin Arlington Robinson
Octaves, XI

Sandage, Allan
What are galaxies? No one knew before 1900. Very few people knew in 1920. All astronomers knew after 1924.

The Hubble Atlas of Galaxies
Galaxies (p. 1)

Shakespeare, William
. . . but when he performs astronomers foretell it.

Troilus and Cressida
Act V, Scene I

Shapley, Harlow
Upton, Winslow
His knees should bend and his neck should curl,
His back should twist and his face should scowl,
One eye should squint and the other protrude,
And this should be his customary attitude.

Popular Astronomy
Harvard Observatory Pinafore (pp. 125–7)
Volume 38, Number 3, March 1930

Shelley, Percy Bysshe
Heaven's utmost deep
Gives up her stars, and like a flock of sheep
They pass before his eye, are numbered, and roll on.

Complete Poetical Works of Shelley
Prometheus Unbound
Act IV, l. 418–20

Silk, Joseph
To many astronomers, the search for intergalactic matter resembles the quest for the holy grail.

Cosmic Enigmas (p. 177)

Swift, Jonathan
This load-stone is under the care of certain astronomers, who from time to time give it such positions as the monarch directs. They spend the greatest part of their lives in observing the celestial bodies, which they do by the assistance of glasses, far excelling ours in goodness.

Gulliver's Travels
Part III
A Voyage to Laputa
Chapter III

There was an astronomer who had undertaken to place a sun-dial upon the great weather-cock on the town-house, by adjusting the annual and diurnal motions of the earth and sun, so as to answer and coincide with all accidental turnings of the wind.

Gulliver's Travels
A Voyage to Balnibarbi
Chapter V

Thoreau, Henry David
One is glad to hear that the naked eye still retains some importance in the estimation of astronomers.

Journal
Volume II, July 8, 1851 (p. 294)

The astronomer is as blind to the significant phenomena, or the significance of phenomena, as the wood-sawyer who wears glasses to defend his eyes from the sawdust.

Journal
Volume II, August 5, 1851 (p. 373)

Unknown
The astronomers must be very clever to have found out the names of all the stars.

The Physics Teacher
A View of the Heavens (p. 413)
Volume 8, Number 7, October 1970

Whitman, Walt
When I heard the learn'd astronomer,
When the proofs, the figures, were ranged
   in columns before me,
When I was shown the charts and diagrams
   to add, divide, and measure them,
When I sitting heard the astronomer where he
   lectured with much applause in the
   lecture-room,
How soon unaccountable I became tired and sick . . .

Complete Poetry and Collected Prose
When I Heard the Learn'd Astronomer

Wordsworth, William
Eyes of some men travel far
For the finding of a star;
Up and down the heavens they go . . .
Like a sage astronomer.

The Poetical Works of William Wordsworth
To the Small Celandine, l. 16–19

Spirits that crowd the intellectual sphere
With mazy boundaries, as the astronomer
With orb and cycle girds the starry throng.

The Poetical Works of William Wordsworth
Ecclesiastical Sonnets
Part II, Section 5, l. 12–14

Young, Edward
Devotion Daughter of Astronomy
An *undevout* astronomer is mad!

Night Thoughts
Night, IX, l. 772–3

. . . *Stars* malign,
Which make their fond *Astronomer* run mad . . .

Night Thoughts
Night IX, l. 1651

ASTRONOMY

Arago, François
Astronomy is a happy science, it has not need for decorations.

Quoted by L.I. Ponomarev in
The Quantum Dice (p. 225)

Brecht, Bertolt
... the new astronomy is a framework of guesses or very little more—yet.

Galileo
Scene One

Comte, Auguste
We may therefore define Astronomy as the science by which we discover the laws of the geometrical and mechanical phenomena presented by the heavenly bodies.

The Positive Philosophy
Volume I
Book II, Chapter I (p. 115)

de Fontenelle, Bernard
... astronomy is the offspring of idleness ...

Conversation on the Plurality of Worlds
First Evening (p. 12)

Dick, Thomas
The objects which astronomy discloses afford subjects of sublime contemplation, and tend to elevate the soul above vicious passions and groveling pursuits.

In Elijah H. Burritt's
Geography of the Heavens
Introduction

Astronomy is that department of knowledge which has for its object to investigate the motions, the magnitudes, and distances of the heavenly bodies; the laws by which their movements are directed, and the ends they are intended to subserve in the fabric of the universe. This is a science which has in all ages engaged the attention of the poet, the philosopher, and the divine, and has been the subject of their study and admiration. Kings have descended from their thrones to render it homage, and have sometimes enriched it with their labours; and humble shepherds, while watching their flocks by night, have beheld with rapture the blue vault of heaven, with its thousand shining orbs, moving in silent grandeur, till the morning star announced the approach of day.

The Works of Thomas Dick, LL.D.
Volume VII
Celestial Scenery
Introduction (p. 8)

Dickson, Frank
The little boy had received his first lesson in astronomy and was proudly exhibiting his knowledge to his still smaller sister.

"That star," he said, pointing to one of the most brilliant ornaments of the heavens, "is much larger than the earth."

"You can't make me believe that," the sister replied. "If it's as big as that, why doesn't it keep the rain off us?"

Quote
Good Stories You Can Use (p. 14)
Volume 52, Number 3, July 17, 1966

Donne, John
We thinke the heavens enjoy their Spherical,
Their round proportion embracing all.
But yet their various and perplexed course,
Observ'd in divers ages doth enforce
Men to finde out so many Eccentrique parts,
Such divers downe-right lines, such overthwarts,
As disproportion that pure forme. It teares
The Firmament in eight and forty sneers . . .

The Complete English Poems of John Donne
The First Anniversary
1. 251–8

Emerson, Ralph Waldo
The first steps in Agriculture, Astronomy . . . teach that Nature's dice are always loaded . . .

The Works of Ralph Waldo Emerson
Volume I
Nature
Discipline (p. 38)

No one can read the history of astronomy without perceiving that Copernicus, Newton, Laplace, are not new men, or a new kind of men, but that Thales, Anaximenes, Hipparchus, Empedocles, Aristorchus, Pythagoras, Oenipodes, had anticipated them . . .

The Works of Ralph Waldo Emerson
Volume VI
Conduct of Life
Fate (p. 18)

Astronomy taught us our insignificance in Nature . . .

The Works of Ralph Waldo Emerson
Volume X
Lectures and Biographical Sketches
Historic Notes of Life and Letters in New England (p. 317)

Esar, Evan
[Astronomy] A great science by which man learns how small he is.

Esar's Comic Dictionary

[Astronomy] A subject that's beyond the depth of some students and over the head of all.

Esar's Comic Dictionary

Flammarion, Camille
Far from being a difficult and inaccessible science, Astronomy is the science which concerns us most, the one most necessary for our general instruction, and at the same time the one which offers for our study the greatest charms and keeps in reserve the highest enjoyments. We cannot be indifferent to it, for it alone teaches us where we are and what we are; and, moreover, it need not bristle with figures, as some severe *savants* would wish us to believe. The algebraical formulæ are merely scaffoldings analogous to those which are used to construct an admirably designed palace. The figures drop off, and the palace of Urania shines in the azure, displaying to our wondering eyes all its grandeur and all its magnificence.

Popular Astronomy: A General Description of the Heavens (p. 1)

Frost, Robert
"But Cygnus isn't in the Zodiac,"
Dick longed to say, but wasn't sure enough
Of his astronomy . . .

The Poetry of Robert Frost
From Plane to Plane
l. 109–11

Hardy, Thomas
The vastness of the field of astronomy reduces every terrestrial thing to atomic dimensions.

Two on a Tower
IV

Hoyle, Fred
If I were asked to define theoretical astronomy in one sentence I should say that it consists of discovering the properties of matter, partly by experiments carried out on the Earth and partly through the detailed observation of near-by space, and in then applying the results to the Universe as a whole.

The Nature of the Universe
The Earth and Near-By Space (p. 5)

Hubble, Edwin
The history of astronomy is a history of receding horizons.

Quoted by D.W. Sciama in
The Unity of the Universe (p. 74)

Huxley, Thomas
When Astronomy was young "the morning stars sang together for joy," and the planets were guided in their courses by celestial hands.

Darwina
The Origin of Species (p. 56)

Iannelli, Richard
The science of celestial bodies, making use of measurements and figures almost as incomprehensible to the average human being as the defense budget.

The Devil's New Dictionary

Jeffers, Robinson
There is nothing like astronomy to pull the stuff out of
    man.
His stupid dreams and red-rooster importance: let him
    count the star-swirls.

The Beginning and the End
Star-Swirls

Kepler, Johannes
Astronomy has two ends, to save the appearances and to contemplate the true form of the edifice of the world.

Quoted by Michael Zeilik in
Astronomy: The Evolving Universe (p. 43)

I am much occupied with the investigation of physical causes. My aim in this is to show that the celestial machine is to be likened not to a divine organism, but rather a clockwork . . .

Quoted by Michael Zeilik in
Astronomy: The Evolving Universe (p. 50)

Larrabee, Eric
Astronomy was independently discovered by Copernicus and Kepler, who sent the news to each other (de nova stella) by sidereal messenger.

Humor from Harper's
Easy Road to Culture, Sort of (p. 89)

A memorable contest once took place in astronomy between the Red Giants and the White Dwarfs, refereed by Hubble, the inventor of the Red Shift. Ever since, all games in astronomy have been played according to Hoyle.

Humor from Harper's
Easy Road to Culture, Sort of (p. 90)

Leacock, Stephen
Astronomy teaches the correct use of the sun and the planets.

Literary Lapses
A Manual of Education (p. 67)

Lichtenberg, Georg Christoph
Astronomy is perhaps the science whose discoveries owe least to chance, in which human understanding appears in its whole magnitude, and through which man can best learn how small he is.

Lichtenberg: Aphorisms and Letters
Notebook c
Aphorism 23

Neugebauer, Otto
I do not hesitate to assert that I consider astronomy as the most important force in the development of science since its origin sometime around 500 B.C. . . .

The Exact Sciences in Antiquity
Introduction (p. 2)

Oey, Sally
A is for Astronomy, the science of far out
B is for Big Bang, how the cosmos came about
C is for Chandrasekhar, who knew things compact
D is for Dark Matter, whose existence is a fact
E is for Eddington, and matters radiative
F is for Faraday, and wave planes rotative

G is for Galaxies, which fly between voids
H is for Hubble, who knew disks from ellipsoids
I is for Ionization, revealing energy states
J is for Julian Day, for periodic dates
K is for Kepler, and his revolution
L is for Local Group, a galaxian profusion
M is for Molecular Cloud, a protostellar batter
N is for Neutron Star, the densest of matter
O is for Oort Cloud, that beyond Pluto lies
P is for Photon, the coveted prize
Q is for Quasar, the most energetic
R is for Redshift, revealing the kinetic
S is for Supernova, nucleosynthesis site
T is for Telescope, gatherer of light
U is for Ultraviolet, seen only from space
V is for Virial Theorem, an equilibrium case
W is for Wolf-Rayet Star, massive and bright
X is for X-ray, where hot things emit light
Y is the fraction of helium by amassed
Z is for Zenith, the highest and last.

The ABC's of Astronomy

Plato
For everyone, as I think, must see that astronomy compels the soul to look upwards and leads us from this world to another.

The Republic
Book VII, l. 529

Soc. At the Egyptian city of Naucratis, there was a famous old god, whose name was Theuth . . . and he was the inventor of many arts such as arithmetic and calculation and geometry and astronomy . . .

Phaedrus
274

But the race of birds was created out of innocent light-minded men, who, although their minds were directed toward heaven, imagined, in their simplicity, that the clearest demonstration of the things above was to be obtained by sight . . .

Timaeos
Section 91

Raymo, Chet
It is easy to be overawed by the visions of the new astronomy. Many among us would prefer to retreat into a comfortable cloud of unknowing. But if we are truly interested in knowing who we are, then we must be

brave enough to accept what our senses and our reason tell us. We must enter into the universe of the galaxies and the light-years, even at the risk of spiritual vertigo, and know what after all must be known.

The Soul of the Night (p. ix)

Shakespeare, William
Doubt thou the stars are fire;
Doubt that the sun doth move;

Hamlet, Prince of Denmark
Act II, scene 2
l. 116–7

Not from the stars do I my judgment pluck,
And yet methinks I have astronomy . . .

Sonnets
XIV

Tennyson, Alfred Lord
We fronted there the learning of all Spain,
All their cosmogonies, their astronomies . . .

Alfred Tennyson's Poetical Works
Columbus
l. 41–2

These are Astronomy and Geology, terrible Muses!

Alfred Tennyson's Poetical Works
Parnassus
Part II, l. 15

Thoreau, Henry David
Astronomy is a fashionable study, patronized by princes, but not fungi.

Journal
Volume XII, October 15, 1859 (p. 390)

Trumbull, John
Though in astronomy survey'd,
His constant course was retrograde;
O'er Newton's system though he sleeps,
And finds his wits in dark eclipse!

Quoted by Florian Cajori in
The Teaching and History of Mathematics in the United States (p. 62)

Unknown
GURU: Today I will discourse upon the violence in astronomy.

DISCIPLE: Revered Sir! Will you be describing the violent phenomena in the Universe?

GURU: Yes, and I will also dwell upon the controversies amongst the astronomers about what these events imply—controversies which are no less violent than the phenomena themselves.

Quoted by Jayant Narlikar in
Violent Phenomena in the Universe (p. 1)

Voltaire
Superstition is to religion what astrology is to astronomy! a very stupid daughter of a very wise mother.

Quoted by J. de Finad in
A Thousand Flashes of French Wit, Wisdom, and Wickedness

Walcott, Derek
I try to forget what happiness was,
and when that don't work, I study the stars.

The Star Apple Kingdom
The Schooner
Flight, Section 11

White, William Hale [Mark Rutherford]
The great beauty of astronomy is not what is incomprehensible in it, but its comprehensibility—its geometrical exactitude.

Quoted by Wilfred Stone in
Religion and Art of William Hale White

Astronomers work always with the past; because light takes time to move from one place to another, they see things as they were, not as they are.

Neale E. Howard – (See p. 5)

ATOMS

Allen, Woody

We can say that the universe consists of a substance, and this substance we call "atoms," or else we call it "monads." Democritus called it atoms. Leibnitz called it monads. Fortunately, the two never met or there would have been a very dull argument.

Getting Even (p. 30)

Bentley, Richard

. . . the *fortuitous or casual concourse of atoms* . . .

The Works of Richard Bentley
Volume III, Sermon vii (p. 147)

Biedny, Demian

The USSR has been labeled the land of the yokel and
    Khamov.
Quite right! And we have an example in this Soviet fellow
    named Gamow.
Why, this working-class bumpkin, this dimwit, this
    Gyorgy Anton'ich called Geo.,
He went and caught up with the atom and kicked it
    about like a pro.

Quoted by George Gamow in
My World Line (p. 74)

Blake, William

The atoms of Democritus,
And Newton's particles of light . . .

BLAKE: The Complete Poems
Mock on Voltaire, Rousseau
l. 9–10

Bohr, Niels
When it comes to atoms, language can be used only as in poetry.

Quoted by K.C. Cole in
Discovery
On Imagining the Unseeable (p. 70)
Volume 3, Number 12, December 1982

The study of atoms . . . not only has deepened our insight into a new domain of experience, but has thrown new light on general problems of knowledge.

Quoted by Robert K. Adair in
The Great Design (p. 194)

Bradley, Omar
We have grasped the mystery of the atom and rejected the Sermon on the Mount.

The Collected Writings of General Omar N. Bradley
Speeches, 1945–1949
Volume 1 (p. 588)

Butler, Samuel
When people talk of atoms obeying fixed laws, they are either ascribing some kind of intelligence and free will to atoms or they are talking nonsense.

Samuel Butler's Notebooks (p. 72)

Atoms have a mind as much smaller and less complex than ours as their bodies are smaller and less complex.

Samuel Butler's Notebooks (p. 73)

The idea of an indivisible atom is inconceivable by the lay mind. If we can conceive an idea of the atom at all, we can conceive it as capable of being cut in half . . .

Samuel Butler's Notebooks (p. 84)

We shall never get people whose time is money to take much interest in atoms.

Samuel Butler's Notebooks (p. 133)

Calder, Ritchie
The atom will always be a sink of energy and never a reservoir.

Profile of Science

Chu, Steven
The atoms become like a moth, seeking out the region of higher laser intensity.

The New York Times
Quoted by James Gleick in
Lasers Slow Atom for Scrutiny (p. 17)
Sunday July 13, 1986
Section A, Column 2

Cicero
The beginnings of all things are small.

Quoted by L.I. Ponomarev in
The Quantum Dice (p. 23)

Fortuitous concourse of atoms.

De Natura Deorum
Book I, Chapter 24, section 66

Cole, A.D.
. . . an atom is a world in itself . . .. How has the indivisible unit evolved into the complex microcosm we now imagine?

Science
Recent Evidence for the Existence of the Nucleus Atom (p. 73)
N.S. Volume 41, Number 1046
Friday January 15, 1915

Dalton, John
These observations have tacitly led to the conclusion which seems universally adopted, that all bodies of sensible magnitude, whether liquid or solid, are constituted of a vast number of extremely small particles, or atoms of matter . . .

A New System of Chemical Philosophy
Chapter 11 (p. 112)

Davies, Paul Charles William
. . . the rules of clockwork might apply to familiar objects such as snookerballs, but when it comes to atoms, the rules are those of roulette.

God and the New Physics
The quantum factor (p. 102)

Democritus of Abdera
Nothing exists except atoms and empty space; everything else is opinion.

Quoted by Leon Lederman in
The God Particle (p. 1)

Dryden, John
From harmony, from heav'nly harmony
     This universal frame began:
  When Nature underneath a heap
     Of jarring atoms lay,
  And could not heave her head . . .

The Poetical Works of Dryden
A Song for St. Cecilia's Day
l. 4–8

The airy atoms did in plagues conspire . . .

The Poetical Works of Dryden
Britannia Rediviva
l. 154

To be the child of chance, and not of care,
No atoms casually together hurl'd.

The Poetical Works of Dryden
To my Honor'd Friend Sir Robert Howard
l. 31–2

So many huddled atoms make a play . . .

The Poetical Works of Dryden
Prologue and Epilogue to the University of Oxford
l. 31

May, tho' our atoms should resolve by chance . . .

The Poetical Works of Dryden
Lucretius
l. 19

For then our atoms, which in order lay,
Are scatter'd from their heap, and puff'd away . . .

The Poetical Works of Dryden
Lucretius
l. 119–20

Durack, J.J.
In the dusty lab'ratory
  'Mid the coils and wax and twine,
There the atoms in their glory
  Ionize and recombine.

The American Physics Teacher
Ions Mine (p. 180)
Volume 7, Number 3, June 1939

Eddington, Sir Arthur Stanley
Our method of making an atom work is to knock it about; and if it does
not do what we want, knock it still harder.

New Pathways in Science (p. 203)

Man is slightly nearer to the atom than to the star . . .. From his central
position he can survey the grandest works of Nature with the astronomer,
or the minutest works with the physicist.

Stars and Atoms
Lecture I (p. 1)

"A fortuitous concourse of atoms"—that bugbear of the theologian—has
a very harmless place in orthodox physics.

The Nature of the Physical World
Chapter IV
Thermodynamical Equilibrium (p. 77)

Einstein, Albert
The unleased power of the atom has changed everything save our modes
of thinking and we thus drift toward unparalleled catastrophe.

The New York Times
Atomic Education Urged by Einstein (p. 13)
Sunday May 25, 1946
Section A, Column 4

Emerson, Ralph Waldo
The intellect sees that every atom carries the whole of Nature . . .

The Works of Ralph Waldo Emerson
Volume VI
The Conduct of Life
Illusions (p. 303)

You cannot detach an atom from its holdings, or strip off from it the
electricity, gravitation, chemic affinity, or the relation to light and heat,
and leave the atom bare.

The Works of Ralph Waldo Emerson
Volume VII
Society and Solitude
Farming (p. 139)

'For the world was built in order,
And the atoms march in tune;
Rhyme the pipe, and Time the warder,
The sun obeys them and the moon.

The Works of Ralph Waldo Emerson
Volume IX
Poems
Monadnoc, l. 245–8

Atom from atom yawns as far
As moon from earth, or star from star.

The Works of Ralph Waldo Emerson
Volume IX
Poems
Nature

Esar, Evan
[Atoms] Another thing that was opened by mistake.

Esar's Comic Dictionary

Faraday, Michael
But I must confess I am jealous of the term atom; for though it is very easy to talk of atoms, it is very difficult to form a clear idea of their nature, especially when compound bodies are under consideration.

Experimental Researches in Electricity
Volume I, 869

Feynman, Richard P.
Atoms are completely impossible from the classical point of view . . .

The Feynman Lectures on Physics
Volume III
Chapter 2-4 (p. 2-6)

Freeman, Ira M.
O'er the atom's wondrous, useful pieces,
    The physicist effuses;
Applauding too as Man releases
    The atom's peaceful uses.

The Physics Teacher
Nuclear Situation Unclear (p. 319)
Volume 13, Number 5, May 1975

Hall, John
If that this thing we call the world
By chance on atoms was begot
Which though in ceaseless motion whirled
Yet weary not
How doth it prove
Thou art so fair and I in love.

Quoted by John D. Barrow in
The World within the World (p. 162)

Hay, Will [Seaton, Ray]
MASTER: They split the atom by firing particles at it, at 5,500 miles a second.

BOY: Good heavens. And they only split it?

Good Morning Boys
The Fourth Form at St. Michael's

Heisenberg, Werner
All the qualities of the atom of modern physics are derived, it has no immediate and direct physical properties at all, i.e., every type of visual conception we might wish to design is *eo ipso*, faulty . . .

Philosophic Problems of Nuclear Science (p. 38)

It is not surprising that our language should be incapable of describing the processes occurring within the atoms, for, as has been remarked, it was invented to describe the experiences of daily life, and these consist only of processes involving exceedingly large numbers of atoms. Furthermore, it is very difficult to modify our language so that it will be able to describe these atomic processes, for words can only describe things of which we can form mental pictures, and this ability, too, is a result of daily experience.

The Physical Principles of the Quantum Theory (p. 11)

Jeans, Sir James Hopwood
The universe was a stage in which always the same actors—the atoms—played their parts, differing in disguises and groupings, but without change of identity. And these actors were endowed with immortality.

The Mysterious Universe
Matter and Radiation (pp. 44–5)

Lamarck, Jean-Baptiste
The most important discoveries of the laws, methods and progress of Nature have nearly always sprung from the examination of the smaller objects which she contains.

Quoted by James R. Newman in
The World of Mathematics
Volume Two (p. 842)

Leibniz, Gottfried Wilhelm
Atoms are the effect of the weakness of our imagination, for it likes to rest and therefore hurries to arrive at a conclusion in subdivisions or analyses; this is not the case in Nature, which comes from the infinite and goes to the infinite. Atoms satisfy only the imagination, but they shock the higher reason.

Quoted by John D. Barrow in
The World within the World (p. 166)

Lemaître, Georges
The atom-world was broken into fragments, each fragment into still smaller pieces. To simplify the matter, supposing that this fragmentation occurred in equal pieces, two hundred sixty generations would have been needed to reach the present pulverization of matter into our poor little atoms, almost too small to be broken again.

The evolution of the world can be compared to a display of fireworks that has just ended: some few red wisps, ashes and smoke. Standing on a well-chilled cinder, we see the slow fading of the suns, and we try to recall the vanished brilliance of the origin of the worlds.

The Primeval Atom
Chapter II (p. 78)

Locke, John
. . . a blind fortuitous concourse of atoms, not guided by an understanding agent . . .

An Essay Concerning Human Understanding
The Expanding Universe
Book IV, Chapter XX, section 15

Maxwell, James Clerk
[Atoms] . . . the imperishable foundation-stones of the universe.

Quoted by James Jeans in
The Mysterious Universe
Matter and Radiation (p. 54)

At quite uncertain times and places,
  The atoms left their heavenly path,
And by fortuitous embraces,
  Engendered all that being hath.

In Lewis Campbell and William Garnett's
The Life of James Clerk Maxwell
Molecular Evolution (p. 637)

How freely he scatters his atoms before the beginning of
  years;
How he clothes them with force as a garment, those small
  incompressible spheres!

In Lewis Campbell and William Garnett's
The Life of James Clerk Maxwell
British Association, 1874 (p. 639)

At any rate the atoms are a very tough lot, and can stand a great deal of knocking about . . .

In Lewis Campbell and William Garnett's
The Life of James Clerk Maxwell
Correspondence (p. 391)

In the very beginning of science,
    the parsons, who managed things then,
Being handy with hammer and chisel,
    made gods in the likeness o men;
Till Commerce arose and at length
    some men of exceptional power
Supplanted both demons and gods by
    the atoms, which last to this hour.

Quoted by John D. Barrow in
The World within the World (p. 168)

Mach, Ernst

If belief in the reality of atoms is so important to you, I cut myself off from the physicist's mode of thinking.

Quoted by Timothy Ferris in
The Red Limit (p. 65)
Letter to Max Planck

Melville, Herman

O Nature, and O soul of Man! how far beyond all utterance are your linked analogies! not the smallest atom stirs or lives in matter, but has its cunning duplicate in mind.

Quoted by John D. Barrow in
The World within the World (p. 1)

Nabokov, Vladimir

But the individual atom is free: it pulsates as it wants, in low or high gear; it decides itself when to absorb and when to radiate energy.

Bend Sinister (p. 159)

Ostwald, Friedrich Wilhelm

We must renounce the hope of representing the physical world by referring natural phenomena to a mechanics of atoms. 'But'—I hear you say—'but what will we have left to give us a picture of reality if we abandon atoms?' To this I reply: 'Thou shalt not take unto thee any graven images, or any likeness of anything.' Our task is not to see the world through a dark and distorted mirror, but directly, so far as the nature of our minds permits. The task of science is to discern relations among realities . . .

Quoted by Nick Herbert in
Quantum Reality (pp. 11–12)

Pope, Alexander
Atoms or systems into ruin hurl'd,
And now a bubble burst, and now a world.

> *The Complete Poetical Works of POPE*
> An Essay on Man
> Epistole I
> Of the Nature and State of Man, with Respect to the Universe
> ·Argument, l. 89–90

See plastic Nature working to this end,
The single atoms each to other tend,
Attract, attracted to, the next in place
Form'd and impell'd its neighbor to embrace.

> *The Complete Poetical Works of POPE*
> An Essay on Man
> Epistle iii, l. 9–12

Robb, Alfred Arthur
All preconceived notions he sets at defiance
By means of some neat and ingenious appliance
By which he discovers a new law of science
Which no one had ever suspected before.
All the chemists went off into fits,
Some of them thought they were losing their wits,
　When quite without warning
　(Their theories scorning)
　The atom one morning
He broke into bits.

> Quoted by Cecilia Helena Payne Gaposchkin in
> *Introduction to Astronomy*
> On J.J. Thomson (p. 341)

What's in an atom, the innermost substratum
That's the problem he is working at today.

> Quoted by Cecilia Helena Payne Gaposchkin in
> *Introduction to Astronomy*
> On Rutherford (p. 341)

A Corpuscle once did oscillate so quickly to and fro,
He always raised disturbances wherever he did go.
He struggled hard for freedom against a powerful foe—
An atom—who wouldn't let him go.

> *The American Physics Teacher*
> The Revolution of the Corpuscle (p. 180)
> Volume 17, Number 3, June 1939

Rutherford, Ernest
[Atom] a nice, hard fellow, red or grey in color according to taste.

Quoted by Robert G. Colodny (Editor) in
From Quarks to Quasars (p. 57)

My work on the atom goes on in fine style. Several atoms succumb each week.

Quoted by Ruth Moore in
Niels Bohr: The Man, His Science, & The World They Changed (p. 114)

Smith, Sydney
Let onion atoms lurk within the bowl . . .

The Wit and Wisdom of Sydney Smith
Recipe for a Salad (p. 429)

Stedman, Edmund Clarence
White orbs like angels pass
   Before the triple glass,
That men may scan the record of each flame,—
      Of spectral line and line
      The legendary divine,—
Finding their mould the same, and aye the same,
   The atoms that we knew before
Of which ourselves are made,—dust, and no more.

The Poems of Edmund Clarence Stedman
Poems of Occasion
Corda Concordia

Stevenson, Adlai
There is no evil in the atom; only in men's souls.

Speech: Hartford Connecticut
18 September 1952

Tennyson, Alfred Lord
If all be atoms, how then should the Gods
Being atomic not be dissoluble,
Not follow the great law?

Poems of Tennyson
Lucretius
l. 114–6

Thompson, Hunter S.
He seemed surprised. "You found a knife that can cut off an atom?" he said. "In this town?"

I nodded. "We're sitting on the main nerve right now," I said.

Quoted by Leon Lederman in
The God Particle (p. 25)

Thomson, William [Lord Kelvin]
The idea of an atom has been so constantly associated with incredible assumptions of infinite strength, absolute rigidity, mystical actions at a distance and indivisibility, that chemists and many other reasonable naturalists of modern times, losing all patience with it, have dismissed it to the realms of metaphysics, and made it smaller than "anything we can conceive".

Nature
On the Size of Atoms (p. 551)
Volume 1, March 31, 1870

Unknown
What's in an atom
The innermost substratum?
That's the problem he is working at today.
He Lately did discover
How to shoot them down the plaver,
And the poor little things can't get away.
He uses as munitions
On his hunting expeditions
Alpha particles which out of Radium sprang.

Quoted by Ruth Moore in
Niels Bohr: The Man, His Science, & The World They Changed (p. 113)

10 October 1929
Rostow na Donu
USSR

Dear Professor Rutherford,

We students of our university physics club elect you our honorary president because you proved that atoms have balls.

Quoted by George Gamow in
My World Line (p. 76)

Vaihinger, Hans
The opponents of the atom are generally content to point to its contradictions and reject it as unfruitful for science. A rash form of caution, for without the atom science falls.

The Philosophy of 'As if' (pp. 70–1)

von Lindermann, Karl Louis Ferdinand
. . . the oxygen atom has the shape of a ring, and the sulphur atom, the shape of a clot.

Quoted by L.I. Ponomarev in
The Quantum Dice (p. 40)

von Weizsacker, Carl Friedrich
. . . only after the atom has lost the last sensible quality does its true meaning for the physical world view become clear; the unity—real, though remote from our immediate perception—of all that our perception knows only as a multitude of appearances is systematically held together and symbolically represented in it, but not mechanically explained.

The World View of Physics (p. 55)

Whitman, Walt
For every atom belongs to me as good as belongs to you.

Complete Poetry and Collected Prose
Song of Myself
I

Wordsworth, William
To let a creed, built in the heart of
    things,
Dissolve before a twinkling atom!

The Poetical Works of William Wordsworth
The Borderers
Act III, l. 1220–1

BIG-BANG

Alphonsus X "The Wise"
. . . if the supreme Being had consulted him when he created the world, he would have given him some good advice.

Quoted by Bernard de Fontenelle in
Conversations on the Plurality of Worlds
First Evening (p. 15)

Čapek, Milič
World history thus began by a "super-radioactive explosion" of the original single quantum, and the development of the universe is a continuation of this process of fragmentation of energy into the increasing number of smaller and smaller quanta. The enormous energy of the cosmic rays is merely a "fossil remnant" of the high-frequency radiation from the original phase of cosmic history.

The Philosophical Impact of Contemporary Physics (p. 352)

Cardenal, Ernesto
And that was Big Bang.
The Great Explosion.
The universe subjected to relations of uncertainty,
its radius of curvature undefined,
   its geometry imprecise
with the uncertainty principle of Quantum Mechanics . . .

Cosmic Canticle
Cantigua 1
Big Bang

Gamow, George
God was very much disappointed, and wanted first to contract the Universe again, and to start all over from the beginning. But it would be much too simple. Thus, being almighty, God decided to correct His mistake in a most impossible way.

31

And God said: "Let there be Hoyle." And there was Hoyle. And God looked at Hoyle . . . and told him to make heavy elements in any way he pleased.

And Hoyle decided to make heavy elements in stars, and to spread them around by supernova explosions.

My World Line (p. 127)

Hoyle, Fred
On scientific grounds this big bang assumption is much the less palatable of the two. For it is an irrational process that cannot be described in scientific terms.

The Nature of the Universe
The Expanding Universe (p. 124)

It is a suspicious feature of the explosion theory that no obvious relics of a superdense state of the Universe can be found.

Frontiers of Astronomy (p. 322)

Jeffers, Robinson
. . . there is no way to express that explosion; all that exists
Roars into flame, the tortured fragments rush away from
  each other into all the sky, new universes
Jewel the black breast of night; and far off the outer nebulae like
  charging spearmen again
Invade emptiness.
      No wonder we are so fascinated with
fire-works.

The Beginning and the End
The Great Explosion

Levi, Primo
Twenty billion years before now,
Brilliant, soaring in space and time
There was a ball of flame, solitary, eternal,
Our common father and our executioner.
It exploded, and every change began.
Even now the thin echo of this one reverse catastrophe
Resounds from the furthest reaches.

Primo Levi Collected Poems
In the Beginning

Silk, Joseph
It's impossible that the Big Bang is wrong.

Quoted by Eric J. Lerner in
The Big Bang Never Happened (p. 11)

Updike, John
Space–time. Three spatial dimensions, plus time. It knots. It freezes. The seed of the universe has come into being. Out of nothing. Out of nothing and brute geometry, laws that can't be otherwise, nobody handed them to Moses, nobody had to. Once you've got that little seed, that little itty-bitty mustard seed—ka-boom! Big Bang is right around the corner.

Roger's Version (p. 303)

Weinberg, Steven
In the beginning there was an explosion. Not an explosion like those familiar on earth, starting from a definite center and spreading out to engulf more and more of the circumambient air, but an explosion which occurred simultaneously everywhere, filling all space from the beginning, with every particle of matter rushing apart from every other particle.

The First Three Minutes
Chapter I (p. 5)

Weldon, Fay
Who cares about half a second after the big bang; what about the half second before?

Quoted by Paul Davies in
About Time (p. 129)

Zel'dovich, Ya.B.
The point of view of a sinner is that the church promises him hell in the future, but cosmology proves that the glowing hell was in the past.

Quoted by Joseph Silk in
The Big Bang (p. 101)

BLACK HOLES

Butthead

Black holes are the bungholes of space.

From Beavis and Butthead television show

Cardenal, Ernesto

But a star a little heavier than a neutron star
is a black hole.
  The fauces of a black hole.
 Like a cosmic vacuum cleaner.
Where gravitation is so great, the curvature so great,
that light is swallowed up.

Cosmic Canticle
Cantiga 3
Autumn Fugue

Chandrasekhar, S.

The black holes of nature are the most perfect macroscopic objects there are in the universe: the only elements in their construction are our concepts of space and time.

Quoted by John D. Barrow in
The World within the World (p. 310)

Eddington, Sir Arthur Stanley

I think there should be a law of Nature to prevent a star from behaving in this absurd way!

Observatory 58
Relativistic Degeneracy (p. 37)

Hawking, Stephen

Although Bekenstein's hypothesis that black holes have a finite entropy requires for its consistency that black holes should radiate thermally, at first it seems a complete miracle that the detailed quantum-mechanical calculations of particle creation should give rise to emission with a thermal spectrum. The explanation is that the emitted particles tunnel out of the black hole from a region of which an external observer has no knowledge other than its mass, angular momentum and electric charge. This means that all combinations or configurations of emitted particles that have the same energy, angular momentum and electric charge are equally probable. Indeed, it is possible that the black hole could emit a television set or the works of Proust in 10 leather-bound volumes . . .

Scientific American
The Quantum Mechanics of Black Holes (p. 40)
Volume 236, Number 1, January 1977

Israel, Werner

It is one of the little ironies of our times that while the layman was being indoctrinated with the stereotype of black holes as the ultimate cookie monsters, the professionals have been swinging round to the almost directly opposing view that black holes, like growing old, are really not so bad when you consider the alternative.

Quoted by John D. Barrow in
The World within the World (p. 312)

Laplace, Pierre Simon

There exist in the heavens therefore dark bodies, as large as and perhaps as numerous as the stars themselves. Rays from a luminous star having the same density as the Earth and a diameter 250 times that of the Sun would not reach us because of its gravitational attraction; it is therefore possible that the largest luminous bodies in the Universe may be invisible for this reason.

Quoted by Jean-Pierre Luminet in
Black Holes (p. 6)

Levi, Primo

The sky is strewn with horrible dead suns,
Dense sediments of mangled atoms.
Only desperate heaviness emanates from them,
Not energy, not messages, not particles, not light.
Light itself falls back down, broken by its own weight.

Levi Primo Collected Poems
The Black Stars

Longair, Malcolm

"Just keep away from the black hole garbage bin by the door as you leave," said the Caterpillar. "It's very useful for getting rid of theoretical papers and weak students!"

Alice and the Space Telescope
Chapter 7 (p. 68)

Ruffini, Remo
Wheeler, John A.
A black hole has no hair.

Proceedings of Conference on Space Physics
Relativistic Cosmology and Space Platforms

No more revolutionary views of man and the universe has one ever been driven to consider seriously than those that come out of pondering the paradox of gravitational collapse, greatest crisis of physics of all time.

Black Holes, Gravitational Waves, and Cosmology (p. 307)

What was once the core of a star no longer visible. The core like a Cheshire cat fades from view. One leaves behind only its grin, the other, only its gravitational attraction. Gravitational attraction, yes; light, no. No more than light do any particles emerge. Moreover, light and particles incident from outside emerge and go down the black hole only to add to its mass and increase its gravitational attraction.

American Scientist
Our Universe: The Known and the Unknown (p. 9)
Volume 56, Number 1, Spring 1968

Thorne, Kip S.
Of all the conceptions of the human mind from unicorns to gargoyles to the hydrogen bomb the most fantastic is the black hole: a hole in space with a definite edge over which anything perhaps can fall and nothing can escape; a hole with a gravitational field so strong that even light is caught and held in its grip; a hole that curves space and warps time.

Cosmology + 1
Chapter 8 (p. 63)

Unknown
A black hole is where God divides by zero.

Source unknown

CALCULATION

Bloch, Felix
Erwin with his psi can do
Calculations quite a few.
But one thing has not been seen
Just what psi really mean.

Quoted by John D. Barrow in
The World within the World (p. 141)

Buck, Pearl S.
"And if hydrogen, what about the hydrogen in sea water? Might not the explosion of the atomic bomb set off an explosion of the ocean itself? Nor was this all that Oppenheimer feared. The nitrogen in the air is also unstable, though less in degree. Might not it, too, be set off by an atomic explosion in the atmosphere?"

"The earth would be vaporized," I said.

"Exactly," Compton said, and with what gravity! "It would be the ultimate catastrophe. Better to accept the slavery of the Nazis than to run the chance of drawing the final curtain on mankind!"

Again Compton took the lead in the final decision. If, after calculation, he said, it were proved that the chances were more than approximately three to one million that the earth would be vaporized by the atomic explosion, he would not proceed with the project. Calculation proved the figures slightly less—and the project continued.

American Weekly
The Bomb—The End of the World?
March 8, 1959

Dirac, Paul Adrien Maurice
I consider that I understand an equation when I can predict the properties of its solutions, without actually solving it.

Quoted by Frank Wilczek and Betsy Devine in
Longing for the Harmonies (p. 102)

Einstein, Albert
If A is success in life, then A equals x plus y plus z. Work is x; y is play; and z is keeping your mouth shut.

Observer
January 1950

Your calculations are correct, but your physics is abominable.

Quoted by A. Berger in
The Big Bang and Georges Lemaître
Monsignor Georges Lemaître (p. 370)

Feynman, Richard P.
Where did we get that [Schrödinger's equation] from? It's not possible to derive it from anything you know. It came out of the mind of Schrödinger
. . .

The Feynman Lectures on Physics
Volume III
Chapter 16-5 (p. 16-12)

Graham, L.A.
Jack be nimble, Jack be quick,
Jack jump over the candlestick,
But figure out β and also time T,
"a" due to gravity, velocity V,
And don't forget $y = VT \sin \beta$
Minus $1/2\, aT^2$, or you'll regret later.
Figure trajectory right to the inch
Or it might be a "singe" instead of a cinch!

Ingenious Mathematical Problems and Methods
Mathematical Nursery Rhyme No. 10

Wittgenstein, Ludwig
The process of *calculating* brings about just this intuition. Calculation is not an experiment.

Tractatus Logico Philosophicus
6.2331 (p. 171)

CAUSE AND EFFECT

Einstein, Albert
In classical mechanics, and no less in the special theory of relativity, there is an inherent epistemological defect which was, perhaps for the first time, clearly pointed out by Ernst Mach . . .. No answer can be admitted as epistemologically satisfactory, unless the reason given is an *observable fact of experience*. The law of causality has not the significance of a statement as to the world of experience, except when *observable facts* ultimately appear as causes and effects.

The Principles of Relativity (pp. 112–3)

Heisenberg, Werner
The chain of cause and effect could be quantitatively verified only if the whole universe were considered as a single system—but then physics has vanished, and only a mathematical scheme remains. The partition of the world into observing and observed system prevents a sharp formulation of the law of cause and effect.

The Physical Principles of the Quantum Theory (p. 58)

Russell, Bertrand
The notion of causality has been greatly modified by the substitution of space–time for space and time . . .. Thus geometry and causation become inextricably intertwined.

The Analysis of Matter
Chapter XXX (p. 313)

CHAOS

Adams, Henry
In plain words, Chaos was the law of nature; Order was the dream of man.

The Education of Henry Adams
The Grammar of Science (p. 451)

Aurelius, Marcus [Antoninus]
Either it is a well arranged universe or a chaos huddled together, but still a universe. But can a certain order subsist in thee, and disorder in the All?

The Meditations of Marcus Aurelius
Book IV, Section 27

Dylan, Bob
Chaos is a friend of mine.

Newsweek
The Two Lives of Bob Dylan
9 December 1985 (p. 93)

Ford, Joseph
Unfortunately, non-chaotic systems are very nearly as scarce as hen's teeth, despite the fact that our physical understanding of nature is largely based upon their study. . . . For centuries, randomness has been deemed a useful, but subservient citizen in a deterministic universe. Algorithmic complexity theory and nonlinear dynamics together establish the fact that determinism actually reigns over a quite finite domain; outside this small haven of order lies a largely uncharted, vast wasteland of chaos where determinism has faded into an ephemeral memory of existence theorems and only randomness survives.

Physics Today
How Random is a Coin Toss?
April 1983 (pp. 46, 44)

Frost, Robert
Let chaos storm!
Let cloud shapes swarm!
I wait for form.

The Poetry of Robert Frost
Pertinax

Harrison, Edward Robert
Perhaps, in the ultimate and unimaginable chaos of a big bang, there lurks the cosmogenic genie who conjures and launches multitudes of universes, each equipped with its own unique laws and fundamental constants . . .

Cosmology, the Science of the Universe (p. 300)

Miller, Henry
The world is what it is and I am what I am . . .. This out there and this in me, all this, everything, the resultant of inexplicable forces. A Chaos whose order is beyond comprehension. Beyond human comprehension.

Black Spring
Third or Fourth Day of Spring (p. 25)

Milton, John
May hope, when everlasting Fate shall yield
To fickle Chance, and *Chaos* judge the strife . . .

Paradise Lost
Book II, l. 232–3

To whom these most adhere,
He rules a moment: Chaos Umpire sits,
And by decision more embroils the fray
By which he reigns . . .

Paradise Lost
Book II, l. 907–10

Nietzsche, Friedrich
You must have chaos in your heart to give birth to a dancing star.

Quoted by Eugene F. Mallove in
The Quickening Universe (p. xiii)

Ovid
Chaos: rudis indigestaque moles . . .
[Chaos: a rough and unordered mass . . .]

Metamorphoses
Book I, l. 7

Pope, Alexander
Not chaos-like, together crush'd and bruis'd,
But, as the world, harmoniously confused:
Where order in variety we see,
And where, tho' all things differ, all agree.

The Complete Poetical Works of POPE
Windsor Forest
1. 13–16

Here she beholds the Chaos dark and deep,
Where nameless somethings in their causes sleep . . .

The Complete Poetical Works of POPE
The Duncaid
Book I, 1. 55–6

Then rose the seed of Chaos, and of Night,
To blot out Order, and extinguish Light . . .

The Complete Poetical Works of POPE
The Duncaid
Book IV, 1. 13–14

Lo! thy dread empire, Chaos! is restor'd;
Light dies before thy uncreating word;
Thy hand great Anarch! lets the curtain
    fall;
And universal Darkness buries all.

The Complete Poetical Works of POPE
The Duncaid
Book IV, 1. 653-6

Shakespeare, William
Chaos is come, again.

Othello, Moor of Venice
Act III, Scene 3, 1. 92

Stewart, Ian
. . . chaos is "lawless behavior governed entirely by law".

Does God Play Dice? (p. 17)

Tsu, Chuang
The ruler of the South Sea was called Light; the ruler of the North Sea, Darkness; and the ruler of the Middle Kingdom, Primal Chaos. From time to time, Light and Darkness met one another in the kingdom of Primal Chaos, who made them welcome. Light and Darkness wanted to repay his kindness and said, "All men have seven openings with which they can see, hear, eat, and breathe, but Primal Chaos has none. Let us

try to give him some." So every day they bored one hole, and on the seventh day, Primal Chaos died.

Inner Chapters
Chapter Seven
The Sage King (p. 166)

His knees should bend and his neck should curl,
His back should twist and his face should scowl,
One eye should squint and the other protrude,
And this should be his customary attitude.

Harlow Shapley and Winslow Upton – (See p. 7)

COMETS

Aristotle
Some of the Italians called Pythagoreans say that the comet is one of the planets, but that it appears at great intervals of time and only rises a little above the horizon.

Meteorology
Book 1, Chapter 6

Coleridge, Samuel Taylor
Guardian and friend of the moon, O Earth, whom the comets
  forget not,
Yea, in the measureless distance wheel around and again they
  behold thee!

The Complete Poetical Works of Samuel Taylor Coleridge
Volume I
Hymn to the Earth

Halley, Edmond
Now we know
The sharply veering ways of comets, once
A source of dread, no longer do we quail
Beneath appearances of bearded stars.

Quoted by Florian Cajori in
*Sir Isaac Newton's Mathematical Principles of Natural Philosophy
and His System of the World* (p. xiv)

Hardy, Thomas
It bends for over Yell'ham Plain,
And we, from Yell'ham Heights,
Stand and regard its fiery train,
So soon to swim from sight.

It will return long years hence, when
As now its strange swift shine

Will fall on Yell'ham; but not then
On that sweet form of thine.

The Collected Poems of Thomas Hardy
The Comet at Yell'ham

Hilderbrand, Wolfgang
Whene'er a comet doth appear,
Come mishap, want, sorrow, and fear;
And never hath a comet's sheen
Without great evil yet been seen.
These dire ill-fortunes do ensue
When a comet appears to view—

Quoted by Bruno H. Burgel in
Astronomy for All (p. 257)

Huygens, Christiaan
But indeed all the whole story of Comets and Planets, and the Production of the World, is founded upon such poor and trifling grounds, that I have often wonder'd how an ingenious man could spend all that pains in making such fancies hang together.

The Celestial Worlds Discover'd (p. 160)

Inscription, Babylonian
A comet arose whose body was bright like the day, while from its luminous body a tail extended, like the sting of a scorpion.

Quoted by Michael Rowan-Robinson in
Our Universe: An Armchair Guide (p. 1)

Kraus, Karl
If the earth had any idea of how afraid the comet is of contact with it.

Half-Truths & One-and-a-Half Truths (p. 109)

The Louisville Journal Editor
That comet is a gay deceiver! He promised to jostle the earth, but has only jilted her. The rogue has told a tale instead of showing one.

Prenticeana (p. 213)

Milton, John
Satan stood
Unterrifi'd, and like a comet burn'd
That fires the length of *Ophiuchus* huge
In th' Arctik sky . . .

Paradise Lost
Book II, l. 707–10

National Geographic Society
Comets are the nearest thing to nothing that anybody can be and still be something.

31 March 1955

Nicholson, Norman
It's here at last. Eyes in the know
Had spotted it two years ago,
A microscopic smut on film.

. . .

Anxious astronomers protest:
Give them a month, and they'll know just what
The frozen core is made of, test
The fluorescence tailing from it,
Fanned out in the solar wind . . .

Collected Poems
Comet Come

Noyes, Alfred
"It was a comet, made of mortal sins; . . .

Watchers of the Sky (p. 61)
Tycho Brahe

Nye, Bill
The comet is a kind of astronomical parody on th ᷈ planet. Comets look some like planets, but they are thinner and do not hurt so hard when they hit anybody as a planet does. The comet was so called because it had hair on it, I believe, but late years the bald-headed comet is giving just as good satisfaction everywhere.

Remarks
Skimming the Milky Way (p. 125)

Paré, Ambroise
The comet was so horrible and frightful . . . that some [people] died of fear and others fell sick. It appeared as a star of excessive length and the color of blood; at its summit was seen the figure of a bent arm holding a great sword in its hand, as if about to strike. On both sides . . . were seen a great number of axes, knives and spaces colored with blood, among which were a great number of hideous human faces with beards and bristling hair.

Quoted by William H. Jefferys and R. Robert Robbins in
Discovering Astronomy
Physician (p. 12)

Seneca
One day there will arise a man who will demonstrate in what region of
the heavens comets take their way.

Quoted by Michael Rowan-Robinson in
Our Universe: An Armchair Guide (p. 1)

Shakespeare, William
When beggars die, there are no comets seen:
The heavens themselves blaze forth the death of princes.

Julius Caesar
Act II, scene ii, l. 30–1

Shelley, Percy Bysshe
Thou too, O Comet, beautiful and fierce,
Who drew the heart of this frail Universe
Towards thine own; till, wrecked in that convulsion,
Alternating attraction and repulsion,
Thine went astray, and that was rent in twain;
Oh, float into our Azure heaven again!

The Complete Poetical Works of Shelley
Epipsychidion
l. 367–72

Swift, Jonathan
. . . the earth very narrowly escaped a brush from the tail of the last
comet, which would have infallibly reduced it to ashes . . .

Gulliver's Travels
Part III, Chapter II

Thomson, James
Lo! from the dread immensity of space
Returning, with accelerated course,
The rushing comet to the sun descends;
And as he shrinks below the shading earth,
With awful train projected o'er the heavens,
The guilty nations tremble.

The Complete Poetical Works of James Thomson
The Seasons
Summer (p. 127)

Turner, H.H.
Of all the meteors in the sky
  There's none like Comet Halley
We see it with the naked eye
  And periodically.

The Mathematical Gazette
Halley's Comet (p. 53)
Volume VI, Number 91, March 1911

Unknown
Far better 'tis, to die
the death that flashes gladness,
than alone, in frigid dignity,
to line on high.
Better, in burning sacrifice,
be thrown against the world
to perish, than the sky
to circle endlessly
a barren stone.

Nature
Nature in Science and Poetry
The Shooting Star (p. 295)
Volume 132, Number 3330, August 26, 1933

. . . obviously, then, comet Kohoutek promises to be the celestial extravaganza of the century.

Newsweek
A Comet for Christmas (p. 109)
November 5, 1973

Whitman, Walt
Year of comets and meteors transient and strange—lo! even
    here one equally transient and strange!
As I flit through you hastily, soon to fall and be gone, what
    is this chant,
What am I myself but one of you meteors?

Complete Poetry and Collected Prose
Year of the Meteor

Young, Edward
Hast thou ne'er seen the Comet's flaming Flight?
Th' illustrious Stranger passing, Terror sheds
On gazing Nations, from his fiery Train
Of length enormous; takes his ample Round
Thro' Depths of Ether; coasts unnumber'd Worlds,
Of more than solar Glory; doubles wide
Heavens's mighty Cape; and then revisits Earth,
From the long Travel of a thousand Years.

Night Thoughts
Night IV, l. 706–13

CONSTELLATIONS

Aeschylus
I know the nightly concourse of the stars
And which of the sky's bright regents brings us storm,
Which summer; when they set, and their uprisings.

Agamemnon
l. 4–6

Aratus
So thought he good to make the stellar groups,
That each by other lying orderly,
They might display their forms. And thus the stars
At once took names and rise familiar now.

Quoted by William Tyler Olcott in
Star Lore of All Ages (p. 3)

Below Orion's feet the Hare
Is chased eternally; behind him
Sirius ever speeds as in pursuit,
And rises after, and eyes him as he sets.

Quoted by Garrett P. Serviss in
Astronomy with the Naked Eye (p. 40)

The tiny Dolphin floats o'er Capricorn,
His middle dusky, but he has four eyes,
Two parallel to two.

Quoted by Garrett P. Serviss in
Astronomy with the Naked Eye (p. 125)

Carlyle, Thomas
Why did not somebody teach me the constellations, and make me at
home in the starry heavens, which are always overhead, and which I
don't half know to this day?

Quoted by Richard Hinckley Allen in
Star Names (p. v)

Coatsworth, Elizabeth
Sweet as violets to a weary heart,
Haunting as the lovely names in old tales,
Beloved as a man's own field, are the Pleiades.

The Pleiades

Frost, Robert
The great Overdog,
That heavenly beast
With a star in one eye,
Gives a leap in the east.

The Poetry of Robert Frost
Canis Major

You know Orion always comes up sideways.

The Poetry of Robert Frost
The Star Splitter
l. 1

Gaposchkin, Cecilia Helena Payne
The *constellations* carry us back to the dawn of astronomy. They have
been called the fossil remains of primitive stellar religion, and as such
they have extraordinary interest.

Introduction to Astronomy (p. 3)

Glasgow, Ellen
Last night the stars were magnificent—Pegasus and Andromeda faced
me brilliantly when I lifted my shade, so I went down and had a friendly
reunion with the constellations—

Letters of Ellen Glasgow
Letter to Mary Johnson (pp. 53–4)
August 15, 1906

Homer
He never closed his eyes, but kept them fixed on the Pleiads, on late-
setting Bootes, and on the Bear—which men also call the wain, and which
turns round and round where it is, facing Orion, and alone never dipping
into the stream of Oceanus—for Calypso had told him to keep this to his
left.

The Odyssey
Book IV, l. 271–6

His armor shone on his breast, like the star of harvest whose rays are
most brilliant among many stars in the murky night; they call it Orion's
dog.

Iliad
xxii, 25–30

Hopkins, Gerard Manley
Now Time's Andromeda on this rock rude,
With not her either beauty's equal or
Her injury's, looks of by both horns of shore
Her flower, her piece of being, doomed dragon's food.

The Poetical Works of Gerard Manley Hopkins
Andromeda, l. 1–4

Kirkup, James
Slung between the homely poplars at the end
of the familiar avenue, the Great
Bear in its lighted hammock swings,
like a neglected gate that neither bars admission nor invites,
hangs on the sagging pole its seven-pointed shape.

Omens of Disaster
Collected Shorter Poems
Volume I
Ursa Major

Longfellow, Henry Wadsworth
Begirt with many a blazing star,
Stood the great giant Algebar,
Orion, the hunter of the beast!

The Poetical Works of Henry Wadsworth Longfellow
The Occultation of Orion

Manilius, Marcus
First Aries, glorious in his golden wool,
Looks back and wonders at the mighty Bull.

Quoted by Mrs. Jesse B. Holman in
The Zodiac, The Constellations and the Heavens (p. 11)

Close by the Kneeling Bull behold
The Charioteer, who gained by skill of old
His name and heaven, as first his steeds he drove
With flying wheels, seen and installed by Jove.

Quoted by Garrett P. Serviss in
Astronomy with the Naked Eye (p. 20)

Bright Scorpio, armed with poisonous tail, prepares
Men's martial minds for violence and wars.
His venom heats and boils their blood to rage
And rapine spreads o'er the unlucky age.

Quoted by Garrett P. Serviss in
Astronomy with the Naked Eye (p. 103)

"First next the Twins, see great *Orion* rise,
His arms extended stretch o'er half the skies;
His stride as large, and with a steady pace
He marches on, and measures a vast space;
On each broad shoulder a bright star display'd,
And three obliquely grace his hanging blade. . ."

> Quoted by Elijah H. Burritt in
> *The Geography of the Heavens* (p. 42)

"And next Bootes comes, whose order'd beams
Present a figure driving of his teams.
Below his girdle, near his knees, he bears
The bright *Arcturus*, fairest of the stars."

> Quoted by Elijah H. Burritt in
> *The Geography of the Heavens* (p. 84)

"Near to Bootes the bright Crown is view'd
And shines with stars of different magnitude:
Or placed in front above the rest displays
A vigorous light, and darts surprising rays.
This shone, since Theseus first his faith betray'd,
The monument of the forsaken maid."

> Quoted by Elijah H. Burritt in
> *The Geography of the Heavens* (p. 95)
> *Astronomicon*
> Book I (p. 15)

Marduk, Babylonian Sun-god
Then Marduk created places for the great gods.
He set up their likeness in the constellations.
He fixed the year and defined its divisions;
Setting up three constellations for each of the twelve months.
When he had defined the days of the year by the constellations,
He set up the station of the zodiac band as a measure of them all,
That none might be too long or too short.

> Quoted by John D. Barrow in
> *The World within the World* (p. 34)

Ovid
Midst golden stars he stands resplendent now,
And thrusts the Scorpion with his bended bow.

> Quoted by Garrett P. Serviss in
> *Astronomy with the Naked Eye* (p. 113)

There is a place above, where Scorpio bent,
In tail and arms surrounds a vast extent;
In a wide circuit of the heavens he shines,
And fills the place of two celestial signs.

Phaeton

Rilke, Rainer Maria
. . . who sets him in a constellation and puts the measuring-stick of
distance in his hands?

The Duino Elegies (p. 25)

Statius
Vast as the starry Serpent, that on high
Tracks the clear ether, and divides the sky,
And southward winding from the Northern Waiu,
Shoots to remoter spheres its glittering train.

Quoted by Elijah H. Burritt in
The Geography of the Heavens (p. 103)

Tennyson, Alfred Lord
Many a night from yonder ivied casement, ere I went to rest,
Did I look on great Orion, sloping slowly to the west.

The Poems of Tennyson
Locksley Hall, l. 7–8

Thomson, James
And fierce Aquarius stains th' inverted year . . .

The Complete Poetical Works of James Thomson
Seasons
Winter

Virgil
But when Astraca's balance, hung on high,
Betwixt the nights and days divides the sky,
Then yoke your oxen, sow your winter grain,
Till cold December comes with driving rain.

Quoted by Elijah H. Burritt in
The Geography of the Heavens (p. 92)

Watts, Isaac

The Ram, the Bull, the Heavenly Twins,
And next the Crab, the Lion shines,
  The Virgin and the Scales;
The Scorpion, Archer, and Sea-goat,
The Damsel with the Watering-pot,
  The Fish with glittering tails.

> Quoted by Cecilia Helena Payne Gaposchkin in
> *Introduction to Astronomy*
> Mnemonic for the zodiacal constellations (p. 4)

White, Henry Kirke

Orion in his Arctic tower . . .

> *Works of Gray, Blair, Beattie, Collins, Thomson and Kirke White*
> Time

Whitman, Walt

I do not want the constellations any nearer,
I know they are very well where they are,
I know they suffice for those who belong to them.

> *Complete Poetry and Collected Prose*
> Song of the Open Road, I

Young, Edward

A *Star* His Dwelling pointed out *below*:
Ye *Pleiades! Arcturus! Mazaroth!*
And thou, *Orion!* of still keener Eye!
Say, ye, who guide the Wilder'd in the Waves,
And bring them out of Tempest into Port!

> *Night Thoughts*
> Night IX, l. 1702–6

COSMOLOGIST

Jeans, Sir James Hopwood
The cosmogonist has finished his task when he has described to the best of his ability the inevitable sequence of changes which constitute the history of the material universe. But the picture which he draws opens questions of the widest interest not only to science, but also to humanity. What is the significance of the vast processes it portrays? What is the meaning, if any there be which is intelligible to us, of the vast accumulations of matter which appear, on our present interpretations of space and time, to have been created only in order that they may destroy themselves?

Astronomy and Cosmogony
Chapter XVII (p. 422)

Zel'dovich, Ya.B.
Cosmologists are often in error, but never in doubt.

Quoted by Rudolf Kippenhahn in
Light from the Depths of Time (p. 1)

COSMOLOGY

Bondi, Sir Hermann
There are probably few features of theoretical cosmology that could not be completely upset and rendered useless by new observational discoveries.

Quoted by G. Borner in
The Early Universe (p. 26)

McCrea, William Hunter
I am always surprised when a young man tells me he wants to work at cosmology; I think of cosmology as something that happens to one, not something one can choose.

Presidential Address to the Royal Astronomical Society
February 1963

Peebles, Phillip James Edwin
In cosmology the reliance on physical simplicity, pure thought and revealed knowledge is carried well beyond the fringe because we have so little else to go on. By this desperate course we have arrived at a few simple pictures of what the Universe may be like. The great goal is now to become more familiar with the Universe, to learn whether any of these pictures may be a reasonable approximation, and if so how the approximation may be improved. The great excitement in cosmology is that the prospects for doing this seem to be excellent.

Physical Cosmology (p. vii)

Rucker, Rudy
I love cosmology: there's something uplifting about viewing the entire universe as a single object with a certain shape. What entity, short of God, could be nobler or worthier of man's attention than the cosmos itself? Forget about interest rates, forget about war and murder, let's talk about *space*.

The Fourth Dimension (p. 91)

Unknown
An elementary particle that does not exist in particle theory should also
not exist in cosmology.

Source unknown

The Cosmos is about the smallest hole that a man can
hide his head in.

Gilbert Keith Chesterton – (See p. 58)

COSMOS

Alfvén, Hannes
To try to write a grand cosmical drama leads necessarily to myth. To try
to let knowledge substitute ignorance in increasingly larger regions of
space and time is science.

Quoted by Eric J. Lerner in
The Big Bang Never Happened (p. 214)

Blagonravov, Anatoly A.
The exploration of the cosmos—the moon and the planets—is a noble
aim. Our generation has the right to be proud of the fact that it has
opened the space era of mankind.

Quoted by M.L. Henry in
Bulletin of the Atomic Scientists
The Lunar Landing and the US–Soviet Equation (p. 29)
Volume 25, Number 7, September 1969

Chaisson, Eric
If we are examples of anything in the cosmos, it is probably of
magnificent mediocrity.

Cosmic Dawn (p. 291)

Chesterton, Gilbert Keith
In the fairy tales the cosmos goes mad, but the hero does not go mad. In
the modern novels the hero is mad before the book begins, and suffers
from the harsh steadiness and cruel sanity of the cosmos.

Quoted by John D. Barrow in
The World within the World (p. 271)

The Cosmos is about the smallest hole that a man can hide his head in.

Orthodoxy (p. 39)

Donne, John
The Sun is lost, and the earth, and no man's wit
Can well direct him where to looke for it.
And freely men confesse that this world's spent,
When in the Planets, and the Firmament
They seeke so many new; then see that this
Is crumbled out againe to his Atomies.
'Tis all in peeces, all cohaerence gone;
All just supply, and all Relation.

The Complete English Poems of John Donne
An Anatomie of the World
First Anniversary
l. 207–14

Sagan, Carl
The size and age of the Cosmos are beyond ordinary human understanding. Lost somewhere between immensity and eternity is our tiny planetary home.

Cosmos (p. 4)

Unsold, Albrecht
The old dream of wireless communication through space has now been realized in an entirely different manner than many had expected. The cosmos' short waves bring us neither the stock market nor jazz from distant worlds. With soft noises they rather tell the physicist of the endless love play between electrons and protons.

Quoted by W.T. Sullivan, III in
Classics in Radio Astronomy
Preface

CRYSTALS

Lonsdale, Dame Kathleen

. . . a crystal is like a class of children arranged for drill, but standing at ease, so that while the class as a whole has regularity both in time and space, each individual child is a little fidgety!

Crystals and X-Rays (p. 22)

DATA

Berkeley, Edmund C.
There is no substitute for honest, thorough, scientific effort to get correct data (no matter how much of it clashes with preconceived ideas). There is no substitute for actually reaching a correct claim of reasoning. Poor data and good reasoning give poor results. Good data and poor reasoning give poor results. Poor data and poor reasoning give rotten results.

> *Computers and Automation*
> Right Answers—A Short Guide for Obtaining Them (p. 20)
> Volume 18, Number 10, September 1969

Lots of people bring you false information.

> *Computers and Automation*
> Right Answers—A Short Guide for Obtaining Them (p. 20)
> Volume 18, Number 10, September 1969

Deming, William Edwards
Anyone can easily misuse good data.

> *Some Theory of Sampling* (p. 18)

There is only one kind of whiskey, but two broad classes of data, good and bad.

> *The American Statistician*
> On the Classification of Statistics (p. 16)
> Volume 2, Number 2, April 1948

Enarson, Harold L.
It does not follow that because something *can* be counted it therefore *should* be counted.

> Speech to Society for College and University Planning, September 1975

Hoyle, Fred
To the optical astronomer, radio data serves like a good dog on a hunt.

> *Galaxies, Nuclei and Quasars* (p. 43)

Koestler, A.

Without the hard little bits of marble which are called "facts" or "data" one cannot compose a mosaic; what matters, however, are not so much the individual bits, but the successive patterns into which you arrange them, then break them up and rearrange them.

The Act of Creation (p. 235)

Mellor, J.W.

By no process of sound reasoning can a conclusion drawn from limited data have more than a limited application.

Higher Mathematics for Students of Chemistry and Physics (p. 4)

Russell, Bertrand

When a man of science speaks of his "data," he knows very well in practice what he means. Certain experiments have been conducted, and have yielded certain observed results, which have been recorded. But when we try to define a "datum" theoretically, the task is not altogether easy. A datum, obviously, must be a fact known by perception. But it is very difficult to arrive at a fact in which there is no element of inference, and yet it would seem improper to call something a "datum" if it involved inferences as well as observation. This constitutes a problem . . .

The Analysis of Matter
Chapter XIX (p. 187)

Unknown

Sint ut sunt aut non sint.
[Accept them as they are or deny their existence.]

Source unknown

If at first you don't succeed, transform your data set.

Source unknown

DOPPLER

Unknown

The easiest way to observe Doppler's effect optically (not acoustically) in one's everyday life is to go out in the evening and look at the cars. Their lights are white or yellow when they approach, but they are red when they are moving away of you.

Source unknown

There was an old German named Brecht
Whose penis was seldom erect.
   When his wife heard him humming
   She knew he was coming—
An example of Doppler effect.

The New Limerick
2575

E = MC²

Chase, Stuart
The Atomic Age is built on Einstein's equation $E = mc^2$, where m is the mass of the atom, and c is the speed of light (186,000 miles per second). You square that, and out of the atom comes quite a bit of energy.

Saturday Review
New Energy for a New Age (p. 14)
Volume 38
January 22, 1955

Unknown
Energy ='milk chocolate square

Source unknown

Wickstrom, Lois
Through the technology of science
it is now possible
to take more mass out of a chamber
than you put into it.

Pandora
$MC^2 = E$

ECLIPSE

Archilochus
Nothing is strange, nothing impossible,
Nor marvelous, since Zeus the father of gods
Brought night to midday when he hid the light
Of the shining sun.

Quoted by Cecilia Helena Payne Gaposchkin in
Introduction to Astronomy (p. 134)

Chinese ode
For the moon to be eclipsed
Is but an ordinary matter.
Now that the sun had been eclipsed
How bad it is!

Quoted by Bertrand Russell in
The ABC of Relativity (p. 35)

Donne, John
How great love is, presence best tryall makes,
But absence tryes how long this love will bee;
To take a latitude,
Sun, or starres, are fitliest view'd
At their brightest, but to conclude
Of Longitudes, what other way have wee,
But to marke when, and where the darke
    eclipses bee?

Quoted by Eric Rogers in
Astronomy for the Inquiring Mind (p. 33)

Hardy, Thomas
Thy shadow, Earth, from Pole to Central Sea,
Now steals along upon the Moon's meek shine
In even monochrome and curving line
Of imperturbable serenity.

Thomas Hardy's Chosen Poems
At a Lunar Eclipse

Pindar's Ninth Paean
Beam of the Sun!
What wilt thou be about, far-seeing one,
O mother of mine eyes, O star supreme,
In time of day
Reft from us? Why, O why has thou perplexed
The might of man,
And wisdom's way,
Rushing forth on a darksome track?

Quoted by Cecilia Helena Payne Gaposchkin in
Introduction to Astronomy
Eclipse and Tides, section 4

Unknown
Working like a carefully rehearsed team, the sun and moon teamed up
yesterday to put on a dramatic celestial show, the plot calling for the
moon to cast its shadow over the sun to create a eclipse.

The Physics Teacher
Et Cetera (p. 79)
Volume 18, Number 1, January 1980

EINSTEIN, ALBERT

Cerf, Bennett
In a notable family called Stein,
There were Gertrude, and Ep, and then Ein.
  Gert's writing was hazy,
  Ep's statues were crazy,
And nobody understood Ein.

Out on a Limerick (p. 76)

Dukas, Helen
Hoffmann, Banesh
At a reception in his [Einstein's] honor at Princeton, when asked to comment on some dubious experiments that conflicted with both relativistic and prerelativistic concepts, he responded with a famous remark—a scientific credo—that was overhead by the American geometer, Professor Oswald Veblen, who must have jotted it down. Years later, in 1930, when Princeton University constructed a special building for mathematics, Veblen requested and received Einstein's permission to have the remark inscribed in marble above the fireplace of the faculty lounge. It was engraved there in the original German: "Raffiniert ist der Herrgott, aber boshaft ist er nicht," which may be translated "God is subtle, but he is not malicious." In his reply to Veblen, Einstein explained that he meant that Nature conceals her secrets by her sublimity and not by trickery.

Albert Einstein, Creator and Rebel (p. 146)

Harrison manuscripts (attributed)
Said Einstein, "I have an equation
which science might call Rabelaisian
Let P be virginity
approaching infinity,
and U be a constant, persuasion.

"Now if P over U be inverted
and the square root of U be inserted
X times over P
The result, Q.E.D.
Is a relative," Einstein asserted.

Quoted by Sachi Sri Kantha in
An Einstein Dictionary (p. 117)

Leacock, Stephen
Einstein himself is not what one would call a handsome man. When seen by members of the Fortnightly Women's Scientific Society in Boston he was pronounced by many of them to be quite insignificant in appearance. . . . Einstein's theories seem to have made a great stir.

Winnowed Wisdom: A New Book of Humor (pp. 25–6)

Marquis, Don
old doc einstein has
abolished time but they
haven t got the news at
sing sing yet

the lives and times of archy & mehitabel
certain maxims of archy

Rossetter, Jack C.
To Einstein, hair and violin,
   we give our final nod;
though understood by just two folks,
   himself—and sometimes—God!

The Mathematics Teacher
Mathematical Notes (p. 341)
Volume XLIII, Number 7, November 1950

Unknown
When Albert Einstein's Theory of Relativity became universally accepted, but before he was internationally famous, he was invited to give many lectures on the topic. One day he came down with a case of laryngitis and was afraid that he would have to cancel a lecture. "No problem," his driver told him, "I've heard your lecture so many times I can repeat it word for word."

As Einstein watched from the back of the room, the driver delivered the speech flawlessly but afterwards was confronted with a question. Without missing a beat the driver replied, "That is 'the' most stupid and elementary question I've ever been asked. I bet even my chauffeur can answer it."

Source unknown

Williams, W.H.
The Einstein and the Eddington
Were counting up their score;
The Einstein's card showed ninety-eight
And Eddington's was more,
And both lay bunkered in the trap
And both stood there and swore.

Quoted by Saghi Sri Kantho in
An Einstein Dictionary (p. 221)

"If I could move faster than light"
Mused Einstein, when a lad so bright;
"I could set off one day,
In a relative way,
and return on the previous night!"

Quoted by Robert L. Weber in
Science with a Smile (p. 101)

To Newton, and to most of the race,
Gravitation is just one special case
Of forces which obey
$F = ma$;
But to Einstein it results from curved space.

Quoted by Robert L. Weber in
Science with a Smile (p. 102)

When Einstein was traveling to lecture in Spain,
He questioned a conductor time and again:
"It may be a while,"
He asked with a smile,
"But when does Madrid reach this train?"

Quoted by Robert L. Weber in
Science with a Smile (p. 102)

A scientist named Lee wrote a note on
A way to change mass into photon.
He showed Einstein his data,
But he made "light of the matter"
And said it was nothing to gloat on!

Quoted by Robert L. Weber in
Science with a Smile (p. 102)

ELECTRONS

Birkeland, Kristian

Space is filled with electrons and flying electric ions of all kinds.

Quoted by Eric J. Lerner in
The Big Bang Never Happened (p. 169)

Bragg, Sir William

. . . an electron springs into existence.

Scientific Monthly
Electrons and Ether Waves (p. 156)
Volume XIV, Number 8, February 1922

Davisson, Clinton

We think we understand the regular reflection of light and x-rays—and we should understand the reflections of electrons as well if electrons were only waves instead of particles . . . It is rather as if one were to see a rabbit climbing a tree, and were to say, "Well that is rather a strange thing for a rabbit to be doing, but after all there is really nothing to get excited about. Cats climb trees—so that if the rabbit were only a cat, we would understand its behavior perfectly."

Quoted by Anthony French and Edwin Taylor in
An Introduction to Quantum Physics (p. 54)

Dingle, H.

He thought he saw electrons swift
  Their charge and mass combine.
He looked again and saw it was
  The cosmic sounding line.
The population then, said he,
  Must be 10^{79}.

Quoted by Sir Arthur Eddington in
The Expanding Universe
The Universe and the Atom (p. 113)
Section IV

Eddington, Sir Arthur Stanley
The electron, as it leaves the atom, crystallises out of Schrödinger's mist like a genie emerging from his bottle.

The Nature of the Physical World
Chapter IX
Relation of Classical Laws to Quantum Laws (p. 199)

. . . an electron would not know how large it ought to be unless there existed independent lengths in space for it to measure itself against.

The Mathematical Theory of Relativity
Chapter V (p. 155)

An electron is no more (and no less) hypothetical than a star.

New Pathways in Science (p. 21)

Einstein, Albert
. . . I should not want to be forced into abandoning strict causality without defending it more strongly than I have so far. I find the idea quite intolerable that an electron exposed to radiation should choose of its own free will not only its moment to jump off, but also its direction. In that case I would rather be a cobbler, or even an employee in a gaming-house, than a physicist.

Quoted by Ronald W. Clark in
Einstein: The Life and Times
Letter to M. Born (p. 211)
April 29, 1924

Feynman, Richard P.
It is the fact that the electrons cannot all get on top of each other that makes tables and everything else solid.

The Feynman Lectures on Physics
Volume III
Chapter 2-4 (p. 2-7)

Fournier d'Albe, E.E.
We may therefore, without in the least interfering with the efficiency of the electron . . . imagine it to be a veritable microcosm.

The Electron Theory (p. 288)

Gamow, George
To keep order and preserve the properties, I never permit more than two electrons to follow the same track; a *ménage à trois* always gives a lot of trouble, you know.

Mr. Tompkins in Paperback
Chapter 10 (p. 115)

Gibson, Charles R.
An electron is a real particle of negative electricity.

Quoted by Frederick Houk Law in
Science in Literature (p. 251)

Harrison, Edward Robert
But electrons do not move in clear-cut orbits like revolving celestial bodies. They dance, and the atom is a ballroom. The electrons perform stately waltzes, weave curvaceous tangos, jitter in spasmodic quicksteps, and rock to frenetic rhythms. They are waves dancing to a choreography composed differently for each kind of atom.

Masks of the Universe (p. 123)

Hoffmann, Banesh
No longer could an electron roam fancy free wherever it wished but, more like a trolley car than a bus, it must keep strictly to the tracks laid down by Bohr . . .

The Strange Story of the Quantum (p. 54)

Jeans, Sir James Hopwood
The electrons may now be pictured as octopus-like structures with tentacles or "tubes of force" sticking out from it in every direction.

Physics and Philosophy
Chapter IV (p. 122)

The hard sphere . . . has always a definite position in space; the electron apparently has not. A hard sphere takes up a very definite amount of room; an electron—well it is probably as meaningless to discuss how much room an electron takes up as it is to discuss how much room a fear, an anxiety, or an uncertainty takes up.

Quoted by Lincoln Barnett in
The Universe and Dr. Einstein (p. 22)

Lodge, Sir Oliver
Electrons have come into existence somehow. The subject of origins usually lies outside science.

Supplement to Nature
The Ether and Electrons (p. 191)
Volume 112, Number 2805, August 4, 1923

Millikan, Robert Andrews
The word "electron" was first suggested in 1891 by Dr. G. Johnstone Stoney as a name for the "natural unit of electricity," namely, that quantity of electricity which must pass though a solution in order to

liberate at one of the electrodes one atom of hydrogen or one atom of any univalent substance.

The Electron
Chapter II (p. 25)

Oppenheimer, J. Robert

If we ask, for instance, whether the position of the electron remains the same, we must say "no"; if we ask whether the electron's position changes with time we must say "no"; if we ask whether it is in motion, we must say "no."

Science and the Common Understanding
A Science in Change (p. 40)

Rutherford, Ernest

It seems to me that you would have to assume that the electron knows beforehand where it is going to stop.

Rutherford at Manchester
Letter to Niels Bohr (p. 127)
March 20, 1913

Standen, Anthony

. . . nothing will do for Mr. Average Citizen but to stuff himself full of electrons, protons, neutrons, neutrinos, genes, chromosomes, glands, hormones, potassium chloride, high-octane gasoline, ultrasonic vibrations, and the theory of relativity.

Science is a Sacred Cow (p. 26)

Unknown

Nature herself does not even know which way the electron is going to go.

Quoted by Richard Feynman in
The Character of Physical Law
Chapter 6 (p. 147)

The Sex Life of an Electron

One night when his charge was pretty high, Micro Henry decided to try to get a cute little coil to let him discharge. He picked up Millie Amp and took her for a ride on his Mega cycle. They rode across the Wheatstone bridge, around by the sine wave, and stopped in a magnetic field by a flowing current.

Micro Henry, attracted by Millie's characteristic curves, soon had her resistance at a minimum and his field was fully excited. He laid her on the ground potential, raised her frequency, lowered her capacitance and pulled out his high voltage probe. He inserted his probe into her test socket, connected them in parallel, and began to short circuit her shunt.

Fully excited Millie Amp exclaimed, "Mho, mho, mho." With his tube operating at its maximum peak current and her coil vibrating from the current flow, she soon reached her maximum peak potential. The excess current flow had gotten her coil hot and Micro Henry started rapidly discharging and soon had drained off every last electron.

They fluxed all night trying various connections and sockets, until Micro Henry's bar magnet had lost all of its field strength. Afterwards, Millie Amp tried self induction and damaged her solenoid. With his batteries fully discharged, Micro Henry was unable to further excite his generator, so they ended it all by reversing polarity and blowing each other's fuses.

Source unknown

ELECTRON: An extremely small particle, so tiny that an average-size person could sit on six billion of them and barely know it.

In Richard Iannelli's
The Devil's New Dictionary

Whitehead, Alfred North
The electrons seems to be borrowing the character which some people have assigned to the Mahatmas of Tibet.

Science and the Modern World
Chapter II (p. 53)

Wolf, Fred Alan
. . . the electron seems to be aware of its own existence. It interacts with itself like any little self-abusive boy behind locked doors. When it does this it generates infinities—an infinite amount of energy, for example. But when mother-physicist comes home and opens the atomic door and observes the electron, the little angel is peacefully obeying the rules of the universe.

Parallel Universes (pp. 69–70)

ENERGY

Bachelard, Gaston
The laboratory technician has succeeded in implementing by means of the atomic pile the Einsteinian principle of inertia of energy.

Quoted by Paul Schlipp in
Albert Einstein, Philosopher-Scientist
The Philosophic Dialectic of the Concepts of Relativity, V (pp. 577–8)

Bishop, Morris
Come, little lad; come, little lass,
  Your docile creed recite:
"We know that Energy equals Mass
  By the Square of the Speed of Light."

A Bowl of Bishop
$E = mc^2$

Blake, William
Energy is eternal delight.

The Prophetic Writings of William Blake
The Marriage of Heaven and Hell
The Voice of the Devil

Energy is the only life . . . and Reason is the bound or outward circumference of Energy.

The Prophetic Writings of William Blake
The Marriage of Heaven and Hell
The Voice of the Devil

Dyson, Freeman J.
We do not know how the scientists of the next century will define energy or in what strange jargon they will discuss it. But no matter what language the physicists use they will not come into contradiction with Blake. Energy will remain in some sense the lord and giver of life, a reality transcending our mathematical descriptions. Its nature lies at the

75

heart of the mystery of our existence as animate beings in an inanimate universe.

Scientific American
Energy in the Universe (p. 51)
Volume 225, Number 3, September 1971

Einstein, Albert
Energy has mass and mass represents energy.

The Evolution of Physics (p. 208)

Feynman, Richard P.
One of the most impressive discoveries was the origin of the energy of the stars, that makes them continue to burn. One of the men who discovered this was out with his girl friend the night after he realized that nuclear reactions must be going on in the stars in order to make them shine. She said "Look at how pretty the stars shine!" He said "Yes, and right now I am the only man in the world who knows why they shine." She merely laughed at him. She was not impressed with being out with the only man who, at that moment, knew why stars shine. Well, it is sad to be alone, but that is the way it is in this world.

The Feynman Lectures on Physics
Volume 1
Chapter 3-4 (p. 3-7)

Gamow, George
Atome prreemorrdiale!
All-containeeng Atome!
Deesolved eento fragments exceedeengly small.
Galaxies formeeng,
Each wiz prrimal enerrgy!

Mr. Tompkins in Paperback
Chapter 6 (p. 57)

Hammond, Allen Lee
Metz, William D.
A point about solar energy that government planners seem to have trouble grasping is that it is fundamentally different from other energy sources. Solar energy is democratic. It falls on everyone and can be put to use by individuals and small groups of people. The public enthusiasm for solar is perhaps as much a reflection for this unusual accessibility as it is a vote for the environmental kindliness and inherent renewability of energy from the sun.

Science
Solar Energy Research: Making Solar After the Nuclear Model? (p. 241)
Volume 197, Number 4300, July 15, 1977

Lemaître, Georges
If we go back in the course of time we must find fewer and fewer quanta, until we find all the energy of the universe packed in a few or even in a unique quantum.

Nature
The Beginning of the World from the Point of View of Quantum Theory (p. 706)
Volume 127, May 9, 1931

Porter, T.C.
Old Dr. Joule he made this rule:
  The self-same energee
Which lifts a gram of matter to
  42640 c. (centimeters)
Will heat a gram of water through
  One centigrade degree.

Radio Times
February 1933

Russell, Bertrand
As a proposition of linguistics: "Energy" is the name of the mathematical expression in question . . .. As a proposition of psychology: our senses are such that we notice what is roughly the mathematical expression in question, and we are led nearer and nearer to it as we refine upon our crude perceptions by scientific observation.

The ABC of Relativity (p. 113)

Rutherford, Ernest
The energy produced by the breaking down of the atom is a very poor kind of thing. Anyone who expects a source of power from the transformation of these atoms is talking moonshine.

The New York Herald Tribune
Atom Powered World Absurd, Scientists Told
September 12, 1933

Stallo, John Bernard
In a general sense, this doctrine [conservation of energy] is coeval with the dawn of human intelligence. It is nothing more than an application of the simple principle that nothing can come from or to nothing . . .

The Concepts and Theories of Modern Physics (pp. 68–9)

ERROR

Unknown

A promising Ph.D. candidate was presenting his thesis at his final examination. He proceeded with a derivation and ended up with something like:

$$F = -MA$$

He was embarrassed, his supervising professor was embarrassed, and the rest of the committee was embarrassed. The student coughed nervously and said "I seem to have made a slight error back there somewhere."

One of the mathematicians on the committee replied dryly, "Either that or an odd number of them!"

Quoted on the Internet

ETHER

Anaxagoras
The formation of the world began with a vortex, formed out of chaos by Energy. This vortex started at the center and gradually spread. It separated matter into two regions, the rare, hot, dry and light material, the aether, in the outer regions, and the heavier, cooler, moist material, the air, in the inner regions. The air condensed in the center of the vortex, and out of the air, the clouds, water and earth separated. But after the formation of earth, because of the growing violence of the rotary motion, the surrounding fiery aether tore stones away from the earth and kindled them to stars, just as stones in a whirlpool rush outward more than water. The sun, moon and all the stars are stones on fire, which are moved round by the revolution of the aether.

Quoted by Walter R. Fuchs in
Mathematics for the Modern Mind (p. 60)

Bruncken, Herbert Gerhardt
Eddington, Einstein, and Jeans one night
  Sailed off on an ether wave,
Sailed on a curve of celestial light
  Into the cosmic cave.

The Physics Teacher
A Space–Time Lullaby (p. 47)
Volume 1, Number 1, April 1963

Eddington, Sir Arthur Stanley
As far as and beyond the remotest stars the world is filled with aether. It permeates the interstices of the atoms. Aether is everywhere . . .. There is no space without aether, and no aether which does not occupy space.

New Pathways in Science (pp. 38–9)

Einstein, Albert
We may sum up as follows: According to the general theory of relativity space is endowed with physical qualities; in this sense, therefore, an ether exists. Space without an ether is inconceivable.

The World as I See It (p. 204)

Hoffmann, Banesh
First we had the luminiferous ether.
Then we had the electromagnetic ether.
And now we haven't e(i)ther.

The Strange Story of the Quantum (p. 33)

Lodge, Sir Oliver
All pieces of matter and all particles are connected together by the ether and by nothing else. In it they move freely and of it they may be composed. We must study the kind of connection between matter and ether. The particles embedded in the ether are not independent of it, they are closely connected with it, it is probable that they are formed out of it: they are not like grains of sand suspended in water, they seem more like minute crystals formed in a mother liquor.

Ether and Reality

The first thing to realise about the ether is its absolute continuity. A deep-sea fish has probably no means of apprehending the existence of water; it is too uniformly immersed in it: and that is our condition in regard to ether.

Ether and Reality

Maxwell, James Clerk
Ethers were invented for the planets to swim in, to constitute electric atmospheres and magnetic effluvia, to convey sensations from one part of our body to another, till all space was filled several times over with ether.

Quoted by Sir James Jeans
The Mysterious Universe
Relativity and Ether (p. 81)

Planck, Max
The ether, this child of sorrow of classical mechanics . . .

Quoted by Jean-Pierre Luminet in
Black Holes (p. 18)

Thomson, Sir J.J.
In fact, all mass is mass of the ether; all momentum, momentum of the ether; and all kinetic energy, energy of the ether. This view, it should be said, requires the density of the ether to be immensely greater than that of any known substance.

Quoted by Sir Oliver Lodge in
Ether and Reality

"Well that is rather a strange thing for a rabbit to be doing, but after all there is really nothing to get excited about. Cats climb trees—so that if the rabbit were only a cat, we would understand its behavior perfectly."

Clinton Davisson – (See p. 70)

EXPERIMENT

Beveridge, W.I.B.
. . . no one believes an hypothesis except its originator but everyone believes an experiment except the experimenter.

The Art of Scientific Investigation (p. 65)

Boyle, Robert
I neither conclude from one single Experiment, nor are the Experiments I make use of, all made upon one Subject: nor wrest I any Experiment to make it quadrare *with* any preconceiv'd Notion. But on the contrary in all kind of Experiments, and all and every one of those Trials, I make the standards (as I may say) or Touchstones by which I try all my former Notions, whether they hold not in weight or measure and touch, Ec. Foras that Body is no other than a Counterfeit Gold, which wants any one of the Properties of Gold (such as are the Malleableness, Weight, Colour, Fixtness in the Fire, Indissolubleness in Aquafortis, and the like), though it has all the other; so will all those notions be found false and deceitful, that will not undergo all the Trials and Tests made of them by Experiments. And therefore such as will not come up to the desired Apex of Perfection, I rather wholly reject and take a new, than by piercing and patching endeavour to retain the old, as knowing old things to be rather made worse by mending than better.

Quoted by Michael Roberts and E.R. Thomas in
Newton and the Origin of Colours (p. 53)

da Vinci, Leonardo
Experiment is the interpreter of nature. Experiments never deceive. It is our judgment which sometimes deceives itself because it expects results which experiment refuses. We must consult experiment, varying the circumstances, until we have deduced general rules, for experiment alone can furnish reliable rules.

Quoted by Oswald Blackwood in
Introductory College Physics (p. 47)

de Fontenelle, Bernard
'Tis no easy matter to be able to make an Experiment with accuracy. The least fact, which offers itself to our consideration, takes in so many other facts, which modify or compose it, that it requires the utmost dexterity to lay open the several branches of its composition, and no less sagacity to find 'em out.

Quoted by Michael Roberts and E.R. Thomas in
Newton and the Origin of Colours (p. 6)

Eddington, Sir Arthur Stanley
. . . he is an incorruptible watch-dog who will not allow anything to pass which is not observationally true.

The Philosophy of Physical Science (p. 112)

Einstein, Albert
The scientific theorist is not to be envied. For Nature, or more precisely experiment, is an inexorable and not very friendly judge of his work. It never says "Yes" to a theory. In the most favorable cases it says "Maybe," and in the great majority of cases simply "No." If an experiment agrees with a theory it means for the latter "Maybe," and if it does not agree it means "No." Probably every theory will someday experience its "No"—most theories, soon after conception.

Quoted by Helen Dukas and Banesh Hoffmann in
Albert Einstein: The Human Side (p. 18)

Heaviside, Oliver
But it is perhaps too much to expect a man to be both the prince of experimentalists and a competent mathematician.

Electromagnetic Theory
Chapter I
Volume I (p. 14)

Maxwell, James Clerk
An Experiment, like every other event which takes place, is a natural phenomenon; but in a Scientific Experiment the circumstances are so arranged that the relations between a particular set of phenomena may be studied to the best advantage. In designing an Experiment the agents and the phenomena to be studied are marked off from all others and regarded as the Field of Investigation.

The Scientific Papers of James Clerk Maxwell (p. 505)

Panofsky, Wolfgang
As experimental techniques have grown from the top of a laboratory bench to the large accelerators of today, the basic components have changed vastly in scale but only little in basic function. More important, the motivation of those engaged in this type of experimentation has hardly changed at all.

Contemporary Physics
Particle Substructure: A Common Theme of Discovery in this Century (p. 23)
Volume 20, Number 1, 1982

Pearson, Karl
It is the old experience that a rude instrument in the hand of a master craftsman will achieve more than the finest tool wielded by the uninspired journeyman.

Life, Letters and Labours of Francis Galton
Volume III (p. 50)

Planck, Max
Experimenters are the shocktroops of science.

Scientific Autobiography (p. 110)

An experiment is a question which science poses to Nature, and a measurement is the recording of Nature's answer.

Scientific Autobiography (p. 110)

Poincaré, Henri
Experiment is the sole source of truth. It alone can teach us something new; it alone can give us certainty.

The Foundations of Science
Science and Hypothesis
Hypothesis in Physics (p. 127)

Robertson, Howard P.
What is needed is a homely experiment which could be carried out in the basement with parts from an old sewing machine and an Ingersol watch, with an old file of *Popular Mechanics* standing by for reference!

Quoted by Paul Arthur Schlipp in
Albert Einstein: Philosopher-Scientist
Geometry as a Branch of Physics (p. 326)

Schrödinger, Erwin
We never experiment with just one electron or atom . . . any more than we can raise Ichthyosauria in the zoo.

British Journal of Philosophical Science
Volume III, Number 11, November 1952

Truesdell, Clifford A.

The hard facts of classical mechanics taught to undergraduates today are, in their present forms, creations of James and John Bernoulli, Euler, Lagrange, and Cauchy, men who never touched a piece of apparatus; their only researches that have been discarded and forgotten are those where they tried to fit theory to experimental data. They did not disregard experiment; the parts of their work that are immortal lie in domains where experience, experimental or more common, was at hand, already partly understood through various special theories, and they abstracted and organized it and them. To warn scientists today not to disregard experiment is like preaching against atheism in church or communism among congressmen. It is cheap rabble-rousing. The danger is all the other way. Such a mass of experimental data on everything pours out of organized research that the young theorist needs some insulation against its disrupting, disorganizing effect. Poincaré said, "The scientist must order; science is made out of facts as a house is made out of stones, but an accumulation of facts is no more science than a heap of stones, a house." Today the houses are buried under an avalanche of rock splinters, and what is called theory is often no more than the trace of some moving fissure on the engulfing wave of rubble. Even in earlier times there are examples. Stokes derived from his theory of fluid friction the formula for the discharge from a circular pipe. Today this classic formula is called the "Hagen–Poiseuille law" because Stokes, after comparing it with measured data and finding it did not fit, withheld publication. The data he had seem to have concerned turbulent flow, and while some experiments that confirm his mathematical discovery had been performed, he did not know of them.

Six Lectures on Modern Natural Philosophy
Method and Taste in Natural Philosophy (pp. 92–3)

Weyl, Hermann

Allow me to express now, once and for all, my deep respect for the work of the experimenter and for his fight to wring significant facts from an inflexible Nature, who says so distinctly "No" and so indistinctly "Yes" to our theories.

The Theory of Groups and Quantum Mechanics (p. xx)

FACT

Bridgman, P.W.
. . . the fact has always been for the physicist the one ultimate thing from which there is no appeal, and in the face of which the only possible attitude is a humility almost religious.

The Logic of Modern Physics (p. 2)

Einstein, Albert
The justification for a physical concept lies exclusively in its clear and unambiguous relation to facts that can be experienced.

Einstein: A Centenary Volume (p. 229)

It seems that the human mind has first to construct forms independently before we can find them in things. Kepler's marvelous achievement is a particularly fine example of the truth that knowledge cannot spring from experience alone, but only from the comparison of the inventions of the mind with observed fact.

Ideas and Opinions
Johannes Kepler (p. 266)

Faraday, Michael
I could trust a fact and always cross-question an assertion.

Quoted by Oswald Blackwood in
Introductory College Physics (p. 413)

France, Anatole [Jean Jacques Brousson]
Less facts! Less facts, if you please, and more figures.

Anatole France Himself
Less Facts

Gardner, Earl Stanley
Facts themselves are meaningless. It's only the interpretation we give those facts which counts.

The Case of the Perjured Parrot (p. 171)

Greenstein, Jesse L.
Knowing how hard it is to collect a fact, you understand why most people want to have some fun analyzing it.

Fortune
Great American Scientists: The Astronomer (p. 149)
Volume 61, Number 5, May 1960

Harrison, Harry
There was an explanation for everything, once you had your facts straight.

Deathworld 1 (p. 88)

Just because you know a thing is true in theory doesn't make it true in fact.

Deathworld 1 (p. 153)

Heinlein, Robert A.
"What are the facts? Again and again and again—what are the facts? Shun wishful thinking, ignore divine revelation, forget what 'the stars foretell,' avoid opinion, care not what the neighbors think, never mind the unguessable 'verdict of history,'—what are the facts, and to how many decimal places? You pilot always into an unknown future; facts are your only clue. Get the facts!"

Time Enough for Love (p. 262)

A fact doesn't have to be understood to be true. Sure, any reasonable mind wants explanations, but it's silly to reject facts that don't fit your philosophy.

Assignment in Eternity
Volume One
Elsewhen (p. 111)

When an apparent fact runs contrary to logic and common sense, it's obvious that you have failed to interpret the fact correctly.

Orphans of the Sky (p. 129)

Huxley, Julian
To speculate without facts is to attempt to enter a house of which one has not the key, by wandering aimlessly round and round, searching the walls and now and then peeping through the windows. Facts are the key.

Essays in Popular Science
Heredity
I, The Behavior of the Chromosomes (pp. 1–2)

Huxley, Thomas
God give me the strength to face a fact though it slay me.

Quoted by George Seldes in
The Great Quotations (p. 344)

Sit down before fact as a little child, follow humbly wherever and whatever abysses nature leads, or you shall learn nothing.

The Life and Letters of Thomas Henry Huxley
Volume I
Letter written on September 23, 1860 (p. 219)

Latham, Peter M.
People in general have no notion of the sort and amount of evidence often needed to prove the simplest fact.

The Collected Works of Dr. P.M. Latham
2:525

Michelson, Albert
The more important fundamental laws and facts of physical science have all been discovered, and these are now so firmly established that the possibility of their ever being supplanted in consequence of new discoveries is exceedingly remote . . . .. Our future discoveries must be looked for in the sixth place of decimals.

Quoted by John D. Barrow in
The World within the World (p. 173)

Planck, Max
Nothing is more interesting to the true theorist than a fact which directly contradicts a theory generally accepted up to that time, for this is his particular work.

A Survey of Physics (pp. 72–3)

Poincaré, Henri
Well, this is one of the characteristics by which we recognize the facts which yield great results. They are those which allow of these happy innovations of language. The crude fact then is often of no great interest; we may point it out many times without having rendered great services to science. It takes value only when a wiser thinker perceives the relation for which it stands, and symbolizes it by a word.

The Foundation of Science
Science and Method
The Future of Mathematics (p. 375)

Science is built up of facts, as a house is built of stones; but an accumulation of facts is no more a science than a heap of stones is a house.

The Foundations of Science
Science and Hypothesis
Hypothesis in Physics (p. 127)

The scientific fact is only the crude fact translated into a convenient language.

The Foundations of Science
The Value of Science
Is Science Artificial? (p. 330)

Szilard, Leo
"I don't intend to publish it; I am merely going to record the facts for the information of God." "Don't you think God knows the facts?" Bethe asked. "Yes," said Szilard. "He knows the facts, but He does not know *this version of the facts.*"

The Collected Works of Leo Szilard: Scientific Papers
Volume I
Preface (p. xix)

Traditional teaching of mathematics
Ignore the facts.

Traditional teaching of physics
Stick to the facts.

Twain, Mark
In the short space of one hundred and seventy six years, the lower Mississippi has shortened itself two hundred and forty two miles. That is an average of a trifle over one mile and a third per year. Therefore, any calm person, who is not blind or idiotic, can see that in the old oolitic silurian period, just a million years ago next November, the Lower Mississippi river was upwards of one million three hundred miles long, and stuck out over the gulf of Mexico like a fishing rod. And by the same token any person can see that seven hundred forty two years from now the Lower Mississippi will be only a mile and three quarters long, and Cairo and New Orleans will have joined their streets together, and be plodding comfortably along under a single mayor and a mutual board of aldermen. There is something fascinating about science. One gets such wholesale returns on conjecture out of such a trifling investment of fact.

Life on the Mississippi
Chapter XVII

Unknown
Never base your argument on a fact: for if the fact is disproved what becomes of the argument.

Quoted by Robert John Strutt in
Life of John William Strutt (fn, p. 270)

Whitehead, Alfred North
. . . irreducible and stubborn facts . . .

Science and the Modern World
Chapter I (p. 4)

The aim of science is to seek the simplest explanation of complex facts. We are apt to fall into the error of thinking that the facts are simple because simplicity is the goal of our quest. The guiding motto in the life of every natural philosopher should be "Seek simplicity and distrust it."

The Concept of Nature (p. 103)

"What are the facts? . . . You pilot always into an unknown future; facts are your only clue. Get the facts!"

Robert A. Heinlein – (See p. 87)

FISSION

Teller, Edward
F stands for fission
That is what things do
When they get wobbly and big
And must split in two.
And just to complete
The atomic confusion,
What fission has done
Can be undone by fusion.

Conversations on the Dark Secrets of Physics (p. 215)

FORCE

Bethe, Hans
The concept of force will continue to be effective and useful.

Source unknown

Faraday, Michael
I do not perceive in any part of space, whether (to use the common phrase) vacant or filled with matter, anything but forces and the lines in which they are exerted.

Quoted by Robert K. Adair in
The Great Design (p. 49)

Graham, L.A.
Hey diddle, diddle,
The cat and the fiddle,
The cow jumped into the blue;
Her leap into action
Took plenty of traction,
The product of Force times mew.

Ingenious Mathematical Problems and Methods
Mathematical Nursery Rhyme No. 8

Humpty Dumpty sat on a wall,
Wondering how hard he would fall.
Force times time, you will agree,
Is equal to mass times velocity.

Ingenious Mathematical Problems and Methods
Mathematical Nursery Rhyme No. 15

Gross, David
One of the best of the many Pauli jokes tells of Pauli's arriving in Heaven and being given, as befits a theoretical physicist, an appointment with God. When granted the customary free wish, he requests that God

explain to him why the value of the fine-structure constant, $\alpha = e^2/(\hbar \times c)$, which measures the strength of the electric force, is 0.00729735 . . .. God goes to the blackboards and starts to write furiously. Pauli watches with pleasure but soon starts shaking his head violently . . .

Physics Today
On the Calculation of the Fine-Structure Constant (p. 9)
Volume 142, Number 13, December 1989

Hembree, Lawrence
Two first graders were standing outside the school one morning.

"Do you think," said one, "that thermo-nuclear projectiles will be affected by radiation belts?"

"No," replied the other, "Once a force enters space . . . "

The school bell rang, "Damn it," said the first kid. "Here we go, back to the old bead stringing.'

Quote
Light Armor (p. 118)
Volume 54, Number 6, April 6, 1967

Mamula, Karl C.
A force is a force, or course, of course,
And we change our speed of our course, of course.
That is, of course, except for a force in a non-inertial frame.

If we want to change our speed or go another way,
We apply a force until we're now where we want to stay.

The Physics Teacher
Physics "Notes" (p. 237)
Volume 13, Number 4, April 1975

Maxwell, James Clerk
An inextensible heavy chain
Lies on a smooth horizontal plane,
An impulsive force is applied at A,
Required the initial motion of K.

In Lewis Campbell and William Garnett's
The Life of James Clerk Maxwell
A Problem in Dynamics (p. 625)

Gin a body meet a body
    Flyin' through the air,
Gin a body hit a body,
    Wil it fly? and where?

In Lewis Campbell and William Garnett's
The Life of James Clerk Maxwell
In memory of Edward Wilson (p. 630)

Force, then is Force, but mark you! not a thing,
> Only a Vector . . .

In Lewis Campbell and William Garnett's
The Life of James Clerk Maxwell
Report on Tait's Lecture on Force (p. 647)

Newton, Sir Isaac

. . . nor could the moon without some such force be retained in its orbit. If this force was too small, it would not sufficiently turn the moon out of a rectilinear course; if it were too great, it would turn it too much, and draw down the moon from its orbit towards the earth.

Mathematical Principles of Natural Philosophy
Definitions, Definition V

The parts of all homogeneal hard bodies which fully touch one another stick together very strongly. And for explaining how this may be, some have invented hooked atoms, which is begging the question; and others tell us that bodies are glued together by rest . . . and others, that they stick together by conspiring motions . . . .. I had rather infer from their cohesion that their particles attract one another by some force, which in immediate contact is exceedingly strong, at small distances performs the chemical operations above mentioned, and reaches not far from the particle with any sensible effect.

Opticks
Book Three, Chapter I (3/4 way through chapter)

Robb, Alfred Arthur

Here's a health to Professor J.J.!
May he hunt for ions for many a day,
> And take observations
> And find the relations
>> Which forces obey.

The American Physics Teacher
Postprandial Proceedings of Cavendish Society
The Don of the Day
Volume 7, Number 3, June 1939

Smoot, George

Using the forces we now know, you can't make the universe we know now.

Quoted by Eric J. Lerner in
The Big Bang Never Happened (p. 32)

Stone, Samuel John
All things are molded by some plastic force
Out of some atoms somewhere up in space.

Harper's Monthly
Soliloquy of a Rationalistic Chicken
September 1875

Winsor, Frederick
Little Miss Muffet
Sits on her tuffet
In a nonchalant sort of way.
With her force field around her
The spider, the dounder,
Is not in the picture today.

The Space Child's Mother Goose (p. 2)

The Galaxy contains more than
100,000,000,000,000,000,000,000,000,000,000,000,000,000.000,000,000
or 100 thousand trillion trillion trillion trillion dust grains.
Neil McAleer – (See p. 97)

FUSION

Unknown

A day without fusion is like a day without sunshine.

Source unknown

Sex is the best form of fusion at room temperature

Source unknown

GALAXIES

Eddington, Sir Arthur Stanley
The running away of the galaxies does not mean that they have a kind of aversion from us.

New Pathways in Science (p. 210)

McAleer, Neil
The Galaxy contains more than
100,000,000,000,000,000,000,000,000,000,000,000,000,000,000,000
or 100 thousand trillion trillion trillion trillion dust grains.

The Mind-Boggling Universe (p. 18)

Tennyson, Alfred Lord
The fires that arch this dusty dot—
Yon myriad worlded-ways—
The vast sun-cluster' gathered blaze,
World-isles in lonely skies,
Whole heavens within themselves amaze
Our brief Humanities.

Alfred Tennyson's Poetical Works
Epilogue, l. 51–6

Updike, John
And beyond our galaxy are other galaxies, in the universe all told at least a hundred billion, each containing a hundred billion stars. Do these figures mean anything to you?

The Centaur (p. 37)

GEOMETRY

Veblen, Oswald

At the same time it will not be forgotten that the physical reality of geometry can not be put in evidence with full clarity unless there is an abstract theory also . . . .. Thus, for example, while the term electron may have more than one physical meaning, it is by no means such a protean object as a point or a triangle.

Science
Geometry and Physics (p. 131)
Volume LVII, Number 1466, February 2, 1923

The branch of physics which is called Elementary Geometry was long ago delivered into the hands of mathematicians for the purposes of instruction. But, while mathematicians are often quite competent in their knowledge of the abstract structure of the subject, they are rarely so in their grasp of its physical meaning.

Science
Geometry and Physics (p. 131)
Volume LVII, Number 1466, February 2, 1923

Wheeler, John A.

There is nothing in the world except curved empty space. Geometry bent one way here describes gravitation. Rippled another way somewhere else it manifests all the qualities of an electromagnetic wave. Excited at still another place, the magic material that is space shows itself as a particle. There is nothing that is foreign and "physical" immersed in space. Everything that is, is constructed out of geometry.

Quoted by Cecil M. DeWitt and John A. Wheeler in
Battelle Rencontres (p. 273)
1967 Lectures in Mathematics and Physics

Whewell, William
This science is one of indispensable use and constant reference, for every student of the laws of nature; for the relations of space and number are the alphabet in which those laws are written. But besides the interest and importance of this kind which geometry possesses, it has a great and peculiar value for all who wish to understand the foundations of human knowledge, and the methods by which it is acquired.

The Philosophy of the Inductive Sciences
Part I, Book 2, Chapter 4, article 8

It does not follow that because something can be counted it therefore should be counted.
Harold L. Enarson – (See p. 61)

GLUONS

Feynman, Richard P.

We call these quanta *gluons*, and say that besides quarks there must be gluons to hold the quarks together.

Quoted by Heinz R. Pagels in
The Cosmic Code (p. 251)

Unknown

O! O! You eight colorful guys
You won't let quarks materialize
You're tricky, but now we realize
You hold together our nucleis.

Source unknown

GOD

Bohm, David
I would put it another way: people had insight in the past about a form of intelligence that had organized the universe and they personalized it and called it "God."

Quoted by Renée Weber in
Dialogues with Scientists and Sages (p. 21)

Bonner, W.B.
It seems to me highly improper to introduce God to solve our scientific problems.

Quoted by Charles-Albert Reichen in
A History of Astronomy (p. 100)

Born, Max
If God has made the world a perfect mechanism, He has at least conceded so much to our imperfect intellects that in order to predict little parts of it, we need not solve innumerable differential equations, but can use dice with fair success.

Quoted by Heinz R. Pagels in
The Cosmic Code (p. 73)

de Condillac, Étienne Bonnot
To the eye of God there are no numbers: seeing all things at one time, he counts nothing.

Source unknown

Dirac, Paul Adrien Maurice
God used beautiful mathematics in creating the world.

Quoted by Heinz R. Pagels in
The Cosmic Code (p. 191)

101

Dostoevsky, Fyodor Mikhailovich
. . . if God exists and if He really did create the world, then, as we all know, He created it according to the geometry of Euclid and the human mind with the conception of only three dimensions in space.

The Brothers Karamozov
Book V, Chapter 3

Einstein, Albert
What I'm really interested in is whether God could have made the world in a different way; that is, whether the necessity of logical simplicity leaves any freedom at all.

Einstein: A Centenary Volume (p. 128)

Erath, V.
God is a child; and when he began to play, he cultivated mathematics. It is the most godly of man's games.

Quoted by Stanley Gudder in
A Mathematical Journey (p. 269)

Erdös, P.
God created two acts of folly. First, He created the Universe in a Big Bang. Second, He was negligent enough to leave behind evidence for this act, in the form of the microwave radiation.

Quoted by John D. Barrow and Frank J. Tipler in
The Anthropic Cosmological Principle (p. 401)

Feynman, Richard P.
God was invented to explain mystery. God is always invented to explain those things that you do not understand. Now, when you finally discover how something works, you get some laws which you're taking away from God; you don't need him anymore. But you need him for the other mysteries. So therefore you leave him to create the universe because we haven't figured that out yet; you need him for understanding those things which you don't believe the laws will explain, such as consciousness, or why you only live to a certain length of time—life and death—stuff like that. God is always associated with those things that you do not understand. Therefore I don't think that the laws can be considered to be like God because they have been figured out.

Quoted by P.C.W. Davies and J. Brown in
Superstrings: A Theory of Everything (p. 208)

Hawking, Stephen
We still believe that the universe should be logical and beautiful; we just dropped the word "God."

Quoted by Renée Weber in
Dialogues with Scientists and Sages (p. 21)

Hill, Thomas
The genuine spirit of Mathesis is devout. No intellectual pursuit more truly leads to profound impressions of the existence and attributes of a Creator, and to a deep sense of our filial relations to him, than the study of these abstract sciences. Who can understand so well how feeble are our conceptions of Almighty Power, as he who has calculated the attraction of the sun and the planets, and weighed in his balance the irresistible force of the lightning? Who can so well understand how confused is our estimate of the Eternal Wisdom, as he who has traced out the secret laws which guide the hosts of heaven, and combine the atoms on earth? Who can so well understand that man is made in the image of his Creator, as he who has sought to frame new laws and conditions to govern imaginary worlds, and found his own thoughts similar to those on which his Creator has acted?

North American Review
The Imagination in Mathematics (pp. 226–7)
Volume 85, Number 176, July 1857

Huxley, Aldous
"Such a really remarkable discovery. I wanted your opinion on it. About God. You know the formula: m over nought equals infinity, m being any positive number? Well, why not reduce the equation to a simpler form by multiplying both sides by nought? In which case, you have m equals infinity times nought. That is to say that a positive number is the product of zero and infinity. Doesn't that demonstrate the creation of the universe by an infinite power out of nothing? Doesn't it . . . "

Point Counter Point
Chapter XI (p. 135)

Infeld, Leopold
Einstein uses his concept of God more often than a Catholic priest.

Quest—An Autobiography (p. 268)

Jacobi, Karl Gustav
God ever arithmetizes.

Quoted by E.T. Bell in
Men of Mathematics (p. xxi)

Kepler, Johannes
The chief aim of all investigations of the external world should be to discover the rational order and harmony which has been imposed on it by God and which He revealed to us in the language of mathematics.

Quoted by Morris Kline in
Mathematical Thought from Ancient to Modern Times (p. 231)

Leibniz, Gottfried Wilhelm
As God calculates, so the world is made.

Quoted by Morris Kline in
Mathematics and the Physical World (p. 385)

Mencken, H.L.
It is impossible to imagine the universe run by a wise, just and omnipotent God, but it is quite easy to imagine it run by a board of gods. If such a board actually exists it operates precisely like the board of a corporation that is losing money.

Minority Report: H.L. Mencken's Notebooks
Sample 79

Pauli, Wolfgang
I cannot believe God is a weak left-hander.

Quoted by Leon Lederman in
The God Particle (p. 256)

Plato
God ever geometrizes.

Quoted by E.T. Bell in
Men of Mathematics (p. xvii)

Popper, Karl R.
The intuitive idea of determinism may be summed up by saying that the world is like a motion-picture film: the picture or still which is just being projected is *the present*. Those parts of the film which have already been shown constitutes *the past*. And those which have not yet been shown constitute *the future*.

In the film, the future co-exists with the past; and the future is fixed, in exactly the same sense as the past. Though the spectator may not know the future, every future event, without exceptions, might in principle be known with certainty, exactly like the past, since it exists in the same sense in which the past exists. In fact, the future will be known to the producer of the film—to the Creator of the world.

The Open Universe (p. 102)

Sagan, Carl
"God" may be thought of as the cosmic watchmaker, the engineer who constructed the initial state and lit the fuse.

In Stephen Hawking's
A Brief History of Time
Introduction (p. x)

Santayana, George
God then becomes a poetic symbol for the material tenderness and the paternal strictness of this wonderful world; the ways of God become the subject-matter of physics.

The Realm of Matter (p. 205)

Shaw, George Bernard
KNELLER: To you the universe is nothing but a clock that an almighty clockmaker has wound up and set going for all eternity.

NEWTON: Shall I tell you a secret, Mr. Beautymonger? The clock does not keep time. If it did there would be no further need for the Clockmaker . . .. Can you, who know everything because you and God are both artists, tell me what is amiss with the perihelion of Mercury?

KNELLER: The what?

NEWTON: The perihelion of Mercury.

KNELLER: I do not know what it is.

NEWTON: I do. But I do not know what is amiss with it. Not until the world finds this out can it do without the Clockmaker in the heavens . . .

In Good King Charles's Golden Days
Act I

Updike, John
"The most miraculous thing is happening. The physicists are getting down to the nitty-gritty, they've really just about pared things down to the ultimate details, and the last thing they ever expected to happen is happening. God is showing through."

"Mr. Kohler, What kind of God is showing through, exactly?"

Roger's Version (p. 9)

Zel'dovich, Ya.B.
. . . almighty God throwing dice for every single proton or antiproton would soon get tired with the astronomical number of particles. He could not make the asymmetry large enough.

Quoted by Joseph Silk in
Cosmic Enigmas (p. 7)

GRAVITY

Bierce, Ambrose
GRAVITATION, *n.* The tendency of all bodies to approach one another with a strength proportioned to the quantity of matter they contain—the quantity of matter they contained being ascertained by the strength of their tendency to approach one another. This is a lovely and edifying illustration of how science, having made A the proof of B, makes B the proof of A.

The Enlarged Devil's Dictionary

One day in the year 1666 Newton had gone to the country, and seeing the fall of an apple, as his niece told me, let himself be led into a deep meditation on the cause which thus draws every object along a line whose extension would pass almost through the center of the Earth.

Quoted by Charles W. Misner *et al* in
Gravitation (p. 3)

Bronowski, Jacob
There are two experiences on which our visual world is based: that gravity is vertical, and that the horizon stands at right angles to it.

The Ascent of Man (p. 157)

Dürrenmatt, Friedrich
My mission is to devote myself to the problems of gravitation, not the physical requirements of a woman.

The Physicists
Act One (p. 19)

Einstein, Albert
I have learned something else from the theory of gravitation: No ever so inclusive collection of empirical facts can ever lead to the setting up of

such complicated equations. A theory can be tested by experience, but there is no way from experience to the setting up of a theory.

Quoted by Paul A. Schlipp (Editor) in
Albert Einstein, Philosopher-Scientist
Autobiographical Notes (p. 89)

I know that hardly any physicists believe that the gravitational forces can play any part in the constitution of matter. The physicist always argues that the forces are too small. This reminds me of a joke. An unmarried woman had a child and the family was greatly humiliated. So the midwife tried to console the mother by saying: "Don't worry so much, it's a very small child!"

Quoted by Leopold Infeld in
Quest—An Autobiography (p. 266)

Emerson, Ralph Waldo
The child amidst his baubles is learning the action of light, motion, gravity . . .

The Works of Ralph Waldo Emerson
Volume I
Address (p. 121)

Footner, Hulbert
Stepping to the edge of the cut bank, they looked over. The precipitous slide of earth, almost as pale as snow at their feet, was gradually swallowed in the murk. The fact that they could not see the bottom of it made the leap appear doubly terrible. "Remember to let yourself go limp when you hit the dirt", he said. "Gravity will do the rest. I'll be there before you, because I'm heavier."

A Backwoods Princess (p. 245)

Fuller, Thomas
Gravity is the ballast of the soul, which keeps the mind steady.

The Holy and Profane States
Of Gravity (p. 258)

Graham, L.A.
Fiddle de dum, fiddle de dee,
A young man sat beneath a tree
And pondered over gravity:
"Cm sub one times m sub two
Divided by d squared, 'tis true,
Denotes the force that follows through."
Then m sub one hit youth on dome
And gave conviction to this pome.

Ingenious Mathematical Problems and Methods
Mathematical Nursery Rhyme No. 26

Heaviside, Oliver
. . . an old idea that the speed of gravitation must be an enormous
multiple of the speed of light . . . is only moonshine.

Electromagnetic Theory
Volume III
Chapter X (p. 144)

King, Alexander
Newton saw an apple fall and discovered the Laws of Gravity.
Eve made an apple fall and discovered the Gravity of Law . . .

I Should Have Kissed Her More (p. 51)

Laplace, Pierre Simon
*. . . we must suppose that the gravitating fluid has a velocity which is at least
a hundred millions of times greater than that of light . . .*

Celestial Mechanics
Volume 4 (p. 645)

Lehman, Robert C.
Archimedes in his bathtub
Thinking of the king's new crown.
Was it gold or baser metal?
On Archy's brow there was a frown.

Gravitation pulled him downward
As the tub began to fill.
Buoyant forces lifted upward
Till the tub began to spill.

The Physics Teacher
Eureka (p. 87)
Volume 21, Number 2, February 1983

Moore, Mary
[A properly fitted corset] prevents gravity from pulling us too far forward
or too far backward, which in so doing, makes us old before our time.

Quoted by Martin Gardner in
Fads and Fallacies (p. 96)

Newton, Sir Isaac
To understand the motions of the planets under the influence of gravity
without knowing the cause of gravity is as good a progress in philosophy
as to understand the frame of a clock and the dependence of the wheels
upon one another without knowing the cause of the gravity of the weight.

Memoirs of Literature
XVIII

You sometimes speak of gravity as essential and inherent to matter. Pray do not ascribe that notion to me; for the cause of gravity is what I do not pretend to know, and therefore would take more time to consider of it.

Quoted by Richard Bentley in
The Works of Richard Bentley
Letters from Sir Isaac Newton
Letter I
Volume 3 (p. 210)

Thomson, James

. . . by the blended power
Of *gravitation* and *projection*, saw
The whole in silent harmony revolve . . .

. . .

And ruled unerring by that single power
Which draws the stone projected to the ground

The Complete Poetical Works of James Thomson
To the Memory of Newton
l. 40–2, l. 75–6

Unknown
What goes up must come down.

Source unknown

Wheeler, John A.
Gravity, you have led us far
From the boundary of a boundary is zero
To momenergy as moment of rotation.

A Journey into Gravity and Spacetime
Chapter 9 (p. 149)

. . . one feels that he has at last in gravitational collapse a phenomenon where general relativity dramatically comes into its own, and where its fiery marriage with quantum physics will be consummated.

Relativity, Groups, and Topology
Geometrodynamics and the Issue of the Final State (p. 518)
New Prospect for a Decisive Test of General Relativity at the Quantum Level

Whewell, William
[The law of gravitation] is indisputably and incomparably the greatest scientific discovery ever made, whether we look at the advance which it involved, the extent of truth disclosed, or the fundamental and satisfactory nature of this truth.

History of the Inductive Sciences
Part I, Book 7, Chapter 2, article 5

Whittaker, Sir Edmund
Gravitation simply represents a continual effort of the universe to straighten itself out.

Quoted by Robert G. Colodny (Editor) in
From Quarks to Quasars: philosophical problems of modern physics (p. 181)

. . . the electron seems to be aware of its own existence. It interacts with itself like any little self-abusive boy behind locked doors.

Fred Alan Wolf – (See p. 74)

HEAT

Baumel, Judith
Think of the complexity
of temperature, quantification
of that elusive quality "heat."
Tonight, for instance,
your hands are colder than mine.
Someone could measure
more precisely than we
the nature of this relationship.

The Weight of Numbers
Fibonacci

Flanders, Michael
Swann, Donald
You can't pass heat from a cooler to a hotter.
Try if you like, you far better notter,
'cause the cold in the cooler will get hotter as a ruler,
'cause the hotter body's heat will pass to the cooler.

At the Drop of Another Hat
The First and Second Law

Heat is work and work's a curse
And all the heat in the universe
Is gonna cool down
Because it can't increase.

At the Drop of Another Hat
The First and Second Law

111

Fourier, [Jean Baptiste] Joseph
Heat, like gravity, penetrates every substance of the universe, its rays occupy all parts of space. The object of our work is to set forth the mathematical laws which this element obeys. The theory of heat will hereafter form one of the most important branches of general physics.

Analytical Theory of Heat
Preliminary Discourse

Frost, Robert
Say something to us we can learn.
By heart and when alone repeat.
Say something! And it says "I burn."
But say with what degree of heat.
Talk Fahrenheit, talk Centigrade.
Use language we can comprehend.
Tell us what elements you blend.

The Poetry of Robert Frost
Take Something like a Star

Hoyle, Fred
Hoyle, Geoffrey
The Yela can destroy the Earth by wrapping a blanket of hydrogen around our atmosphere. Then all it needs do to destroy us is just press a little of the hydrogen into the atmosphere itself. The hydrogen and the oxygen in our atmosphere combine together with an immense release of heat. The generation of heat causes the gas to rise and more hydrogen is sucked down. Within seconds the whole atmosphere is a raging inferno.

Into Deepest Space
Chapter I (p. 7)

Keane, Bill
Heat makes things expand. That's why the days are longer in the summer.

Caption to cartoon

Mayer, Julius Robert von
Concerning the intimate nature of heat, or of electricity, etc., I know nothing, any more than I know the *intimate nature* of any matter whatsoever, or of anything else.

Kleinere Schriften und Briefe (p. 181)
Quoted by Pierre Duhem in
The Aim and Structure of Physical Theory (p. 52)

Metsler, William
The stove is hot, but that's no change
Heat's what it's supposed to make
Resistance generates the energy to bake
So its always Ohm Ohm on the range.

The Physics Teacher
The Cowboy's Lament (p. 127)
Volume 15, Number 2, February 1977

Newton, Sir Isaac
Do not all fixed bodies, when heated beyond a certain degree, emit light and shine; and is not this emission performed by the vibrating motions of their parts?

Opticks
Book III, Part 1, Query 8

Poincaré, Henri
And a well-made language is no indifferent thing; not to go beyond physics, the unknown man who invented the word *heat* devoted many generations to error. Heat has been treated as a substance, simply because it was designated by a substance, and it has been thought indestructible.

The Foundations of Science
The Value of Science
Analysis and Physics (p. 289)

Unknown
Oh Langley devised the bolometer:
It's really a kind of thermometer
Which measures the heat
From a polar bear's feet
At a distance of half a kilometre.

Source unknown

GO AND SEE WHAT HE'S UP TO –
OR ARE YOU GETTING COLD FEET?!

HYPOTHESIS

Asquith, Herbert
Jolie hypothèse quelle explique tant de choses.
[A pretty hypothesis which explains so many things.]

Speech in the House of Commons
March 29, 1917

Baez, Joan
. . . hypothetical questions get hypothetical answers.

Daybreak
What Would You Do If (p. 134)

Barry, Frederick
Hypothesis, however, is an inference based on knowledge which is insufficient to prove its high probability.

The Scientific Habit of Thought
The Elements of Theory (p. 164)

Bruner, Jerome Seymour
The shrewd guess, the fertile hypothesis, the courageous leap to a tentative conclusion—these are the most valuable coin of the thinker at work.

The Process of Education (p. 14)

Carroll, Lewis
"Would you tell me, please, which way I ought to go from here?"

"That depends a good deal on where you want to get to," said the Cat.

"I don't much care where—" said Alice.

"Then it doesn't matter which way you go," said the Cat.

The Complete Works of Lewis Carroll
Alice's Adventures in Wonderland
Pig and Pepper

Cohen, Morris
There is . . . no genuine progress in scientific insight through the Baconian method of accumulating empirical facts without hypotheses or anticipation of nature. Without some guiding idea we do not know what facts to gather . . . we cannot determine what is relevant and what is irrelevant.

A Preface to Logic (p. 148)

Copernicus, Nicolaus
Since the newness of the hypotheses of this work—which sets the earth in motion and puts an immovable sun at the center of the universe—has already received a great deal of publicity, I have no doubt that certain of the savants have taken grave offense and think it wrong to raise any disturbance among liberal disciplines which have had the right set-up for a long time now.

On the Revolutions of the Heavenly Spheres
Introduction

Cort, David
But suspicion is a thing very few people can entertain without letting the hypothesis turn, in their minds, into fact . . .

Social Astonishments
ONE
Believing in Books

Duhem, Pierre
Contemplation of a set of experimental laws does not, therefore, suffice to suggest to the physicist what hypotheses he should choose in order to give a theoretical representation of these laws; it is also necessary that the thoughts habitual with those among whom he lives and the tendencies impressed on his own mind by his previous studies come and guide him, and restrict the excessively great latitude left to his choice by the rules of logic. How many parts of physics retain to this day a merely empirical form until circumstances prepare the genius of a physicist to conceive the hypothesis which will organize them into a theory!

The Aim and Structure of Physical Theory (p. 255)

Du Noüy, Pierre L.
The man of science who cannot formulate a hypothesis is only an accountant of phenomena.

The Road to Reason (p. 77)

Evans, Bergen
An honorable man will not be bullied by a hypothesis.

The Natural History of Nonsense
A Tale of a Tub

We see what we want to see, and observation conforms to hypothesis.

The Natural History of Nonsense
A Tale of a Tub

Fabing, Harold
Marr, Ray
Many confuse hypothesis and theory. An hypothesis is a possible explanation; a theory, the correct one.

Fischerisms (p. 7)

Freud, Sigmund
In the complete absence of any theory of the instincts which would help us to find our bearings, we may be permitted, or rather, it is incumbent upon us, in the first place to work out any hypothesis to its logical conclusion, until it either fails or becomes confirmed.

On Narcissism
I

Goethe, Johann Wolfgang von
Hypotheses are the scaffolds which are erected in front of a building and removed when the building is completed. They are indispensable to the worker; but he must not mistake the scaffolding for the building.

Maxims and Reflections

Holmes, Sherlock
"If the fresh facts which come to our knowledge all fit themselves into the scheme, then our hypothesis may gradually become a solution."

In Arthur Conan Doyle's
The Complete Sherlock Holmes
The Adventure of Wisteria Lodge

Huxley, Thomas
The great tragedy of Science—the slaying of a beautiful hypothesis by an ugly fact.

Collected Essays
Biogenesis and Abiogenesis

. . . it is the first duty of a hypothesis to be intelligible . . .

Man's Place in Nature
II (p. 126)

Laplace, Pierre Simon
Laplace had presented Napoleon with a copy of the work [*Mécanique céleste*]. Thinking to get a rise out of Laplace, Napoleon took him to task for an apparent oversight. "You have written this huge book on the system of the world without once mentioning the author of the universe." "Sire," Laplace retorted, "I have no need of that *hypothesis.*" When Napoleon repeated this to Lagrange, the latter remarked "Ah, but that is a fine hypothesis. *It explains so many things.*"

Quoted by E.T. Bell in
Men of Mathematics (p. 181)

Lewis, C.S.
This is called the inductive method. Hypothesis, my dear young friend, establishes itself by a cumulative process: or, to use popular language, if you make the same guess often enough it ceases to be a guess and becomes a Scientific Fact.

The Pilgrim's Regress: An Allegorical Apology for Christianity,
Reason and Romanticism (p. 37)

Lorenz, Konrad
It is a good morning exercise for a research scientist to discard a pet hypothesis every day before breakfast. It keeps him young.

On Aggression (p. 12)

Newton, Sir Isaac
We are to admit no more cause of natural things than such as are both true and sufficient to explain their appearances.

Mathematical Principles of Natural Philosophy
Book III, Rule I

In experimental philosophy we are to look upon propositions inferred by general induction from phenomena as accurately or very nearly true, notwithstanding any contrary hypotheses that may be imagined, till such time as other phenomena occur, by which they may either be made more accurate, or liable to exceptions.

Mathematical Principles of Natural Philosophy
Book III, Rule IV

I frame no hypotheses; for whatever is not deduced from the phenomena is to be called an hypothesis; and hypotheses, whether metaphysical or physical, whether of occult qualities or mechanical, have no place in experimental philosophy.

Mathematical Principles of Natural Philosophy
Book III, General Scholium

Nordman, Charles

Hypotheses in science are a kind of soft cement which hardens rapidly in the open air, thus enabling us to join together the separate blocks of the structure and to fill up the breaches made in the walls by projectiles, with artificial stuff which the superficial observer presently mistakes for stone. It is because hypotheses are something like that in science that the best scientific theories are those which include the least hypotheses.

Einstein and the Universe (pp. 34–5)

Pascal, Blaise

For sometimes an obvious absurdity follows from its negation, and then the hypothesis is true and certain; or an obvious absurdity follows from its affirmation, and then the hypothesis is considered false; and when we have not yet been able to draw an absurdity either from its negation or from its affirmation, the hypothesis remains doubtful. So that to establish the truth of an hypothesis it is not enough that all the phenomena should follow from it, whereas if there follows from it something opposed to a single phenomenon, that is enough to make certain its falsity.

Scientific Treatises
Concerning the Vacuum
Pascal's Answer to the Reverend Noel

Planck, Max

. . . every hypothesis in physical science has to go through a period of difficult gestation and parturition before it can be brought out into the light of day and handed to others, ready-made in scientific form so that it will be, as it were, fool-proof in the hands of outsiders who wish to apply it.

Where is Science Going? (p. 178)

Poincaré, Henri

The firm determination to submit to experiment is not enough; there are still dangerous hypotheses; first, and above all, those which are tacit and unconscious. Since we make them without knowing it, we are powerless to abandon them.

The Foundations of Science
Science and Hypothesis
Hypothesis in Physics (p. 134)

For a Latin, truth can be expressed only by equations; it must obey laws simple, logical, symmetric and fitted to satisfy minds in love with mathematical elegance.

The Anglo-Saxon to depict a phenomenon will first be engrossed in making a model, and he will make it with common materials, such as our crude, unaided senses show us them . . .. He concludes from the body to the atom.

Both therefore make hypotheses, and this indeed is necessary, since no scientist has ever been able to get on without them. The essential thing is never to make them unconsciously.

The Foundations of Science
Author's Preface to Translation (p. 6)

Sterne, Laurence
It is the nature of an hypothesis, when once a man has conceived it, that it assimilates every thing to itself, as proper nourishment; and, from the first moment of your begetting it, it generally grows stronger by every thing you see, hear, read, or understand. This is of great use.

Tristram Shandy
Book 2, Chapter 19

Unknown
[Hypothesis] Something usually murdered by facts.

Source unknown

INFINITE

Aristotle
A quantity is infinite if it is such that we can always take a part outside what has been already taken.

Physics
Book III, 6

Nature flies from the infinite. For the infinite is unending or imperfect, and Nature ever seeks an end.

Generation of Animals
Book I, 715b15

Bailey, Philip James
God puts his finger in the other scale,
And up we bounce, a bubble. Nought is great
Nor small, with God; for none but he can make
The atom imperceptible, and none
But he can make a world; he counts the orbs,
He counts the atoms of the universe,
And makes both equal; both are infinite.

Festus
IV (p. 83)

Beerbohm, Max
The attempt to conceive Infinity had always been quite arduous enough for me.

Mainly on the Air
A Note on the Einstein Theory (p. 137)

Bell, E.T.
The toughminded suggest that the theory of the infinite elaborated by the great mathematicians of the Nineteenth and Twentieth Centuries, without which mathematical analysis as it is actually used today is impossible, has been committing suicide in an unnecessarily prolonged and complicated manner for the past half century.

Debunking Science

Berkeley, George
Of late the speculations about Infinities have run so high, and grown to such strange notions, as have occasioned no small scruples and disputes among the geometers of the present age. Some there are of great note who, not content with holding that finite lines may be divided into an infinite number of parts, do yet farther maintain that each of those infinitesimals is itself subdivisible into an infinity of other parts or infinitesimals of a second order, and so on ad infinitum. These, I say, assert there are infinitesimals of infinitesimals of infinitesimals, &c., without ever coming to an end; so that according to them an inch does not barely contain an infinite number of parts, but an infinity of an infinity of an infinity ad infinitum of parts.

The Principles of Human Knowledge
Section 130

Blake, William
Too see a World in a grain of sand,
And a Heaven in a wild flower:
Hold Infinity in the palm of your hand,
And Eternity in an hour.

BLAKE: The Complete Poems
The Pickering Manuscript, Auguries of Innocence, l. 1–4

Borges, Jorge Luis
The ignorant suppose that an infinite number of drawings require an infinite amount of time; in reality, it is quite enough that time be infinitely subdivisible, as is the case in the famous parable of the Tortoise and the Hare. This infinitude harmonizes in an admirable manner with the sinuous numbers of Chance and of the Celestial Archetype of the Lottery, adored by the Platonists . . .

Ficciones
The Babylon Lottery (p 70)

Box, G.E.P.
It is a pity, therefore, that the authors have confined their attention to the relatively simple problem of determining the approximate distribution of arbitrary criteria and have failed to produce any sort of justification for

the tests they propose. In addition to those functions studied there are an infinity of others, and unless some principle of selection is introduced we have nothing to look forward to but an infinity of test criteria and an infinity of papers in which they are described.

Journal of the Royal Statistical Society
Discussion (p. 29)
Ser. B., 18, 1956

Camus, Albert
Somebody has to have the last word. Otherwise, every reason can be met with another one and there would never be no end to it.

The Fall (p. 45)

Carlyle, Thomas
The moment of discovery, "spontaneous illumination . . ." The infinite is made to blend itself with the finite, to stand visible, as it were, attainable there.

Quoted by Roger A. MacGowan and Frederick I. Ordway, III in
Intelligence in the Universe (p. 49)

Carus, Paul
Infinity is the land of mathematical hocus pocus. There Zero the magician is king. When Zero divides any number he changes it without regard to its magnitude into the infinitely small; and inversely, when divided by any number he begets the infinitely great.

Monist
The Nature of Logical and Mathematical Thought (p. 69)
Volume 20, Number 1, January 1910

Crane, Hart
But the star-glistered salver of infinity,
The circle, blind crucible of endless space,
Is sluiced by motion,—subjugated never.

The Collected Poems of Hart Crane
The Bridge
Cape Hatteras

de Morgan, Augustus
Great fleas have little fleas upon their backs to bite 'em,
And little fleas have lesser fleas, and so ad infinitum.
And the great fleas themselves, in turn have greater fleas to go on;
While these again have greater still, and greater still, and so on.

Budget of Paradoxes
Volume II
Are Atoms Worlds (p. 191)

Dryden, John
But how can finite grasp Infinity?

The Poetical Works of Dryden
Hind and the Panther
I, l. 105

Eddington, Sir Arthur Stanley
That queer quantity "infinity" is the very mischief, and no rational physicist should have anything to do with it. Perhaps that is why mathematicians represent it by a sign like a love-knot.

New Pathways in Science (p. 217)

Emerson, Ralph Waldo
. . . and thus ever, behind the coarse effect, is a fine cause, which, being narrowly seen, is itself the effect of a finer cause.

The Works of Ralph Waldo Emerson
Volume II
Circles (p. 283)

Froude, James Anthony
Large forms resolve themselves into parts, down so far as we can see into infinity.

Short Studies on Great Subjects
Calvinism (p. 16)

Gleick, James
In the mind's eye, a fractal is a way of seeing infinity.

Chaos (p. 98)

Harrison, Edward Robert
Only a cosmic jester could perpetrate eternity and infinity . . .

Masks of the Universe (p. 201)

If eternity is silliness, then infinity of space is sheer madness.

Masks of the Universe (p. 202)

Hawking, Stephen
In an infinite number universe, every point can be regarded as the center, because every point has an infinite of stars on each side of it.

A Brief History of Time (p. 5)

Hilbert, David
The infinite! No other question has ever moved so profoundly the spirit of man . . .

Quoted by E.T. Bell in
Men of Mathematics (p. xxi)
(Address in memory of Weirstrass)

Huxley, Thomas
Truly it has been said, that to a clear eye the smallest fact is a window through which the Infinite may be seen.

Huxley's Essays
The Study of Zoology (p. 30)

Kant, Immanuel
But the infinite is absolutely (not merely comparatively) great. In comparison with this all else (in the way of magnitudes of the same order) is small. But the point of capital importance is that the mere ability even to think it as a *whole* indicates a faculty of mind transcending every standard sense. For the latter would entail a comprehension yielding as unit a standard bearing to the infinite ration expressible in numbers, which is impossible.

The Critique of Judgment
First Part
The Mathematically Sublime, §26

Kasner, Edward
Newman, James R.
With the Hottentots, infinity begins at three.

Mathematics and the Imagination (p. 19)

Lucretius
Again if for the moment all existing space be held to be bounded, supposing a man runs forward to its outside borders, and stands on the utmost verge and then throws a winged javelin, do you choose that when hurled with vigorous force it shall advance to the point to which it has been sent and fly to a distance, or do you decide that something can get in its way and stop it? for you must admit and adopt one of the two suppositions; either of which shuts you out from all escape and compels you to grant that the universe stretches without end.

On the Nature of Things
Book I, l. 968

Pascal, Blaise
It is an infinite sphere, the centre of which is everywhere, the circumference nowhere.

Pensées
Section II, 72

We know that there is an infinite, and we are ignorant of its nature.

Pensées
Section III, 233

Unity joined to infinity adds nothing to it, no more than one foot to an infinite measure. The finite is annihilated in the presence of the infinite, and becomes a pure nothing.

Pensées
Section III, 233

Phrase, Latin
Ad infinitum
[To infinity]

Richardson, Lewis
Big whorls have little whorls
Which feed on their velocity,
And little whorls have lesser whorls,
And so on to viscosity.

Quoted by Ian Stewart in
Does God Play Dice? (p. 196)

Royce, Josiah
. . . let us suppose, if you please, that a portion of the surface of England is very perfectly leveled and smoothed, and is then devoted to the production of our precise map of England. That in general, then, should be found upon the surface of England, map constructions which more or less roughly represent the whole of England,—all this has nothing puzzling about it . . . .. But now suppose that this our resemblance is to be made absolutely exact . . . .. A map of England, contained within England, is to represent, down to the minutest detail, every contour and marking, natural or artificial, that occurs upon the surface of England . . .

One who, with absolute exactness of perception, looked down upon the ideal map thus supposed to be constructed, would see lying upon the surface of England, and at a definite place thereon, a representation of England on as large or small a scale as you please . . . . This representation, which would repeat in the outer portions the details of the former, but upon a smaller space, would be seen to contain yet another England, and this another, and so on without limit.

The World and the Individual
Supplementary Essay
Section III

Shakespeare, William
. . . I could be bounded in a nutshell and count myself a king of infinite space . . .

Hamlet, Prince of Denmark
Act II, scene 2, l. 263

Shelley, Percy Bysshe
. . . infinity within,
Infinity without . . .

The Complete Poetical Works of Shelley
Queen Mab, l. 22–3

Swift, Jonathan
So, Nat'ralists observe, a Flea
Hath smaller Fleas that on him prey,
And these have smaller Fleas to bite 'em
And so proceed, *ad infinitum*.

The Portable Swift
On Poetry—A Rhapsody

Thomson, William [Lord Kelvin]
I say finitude is incomprehensible, the infinite in the universe is
comprehensible. Now apply a little logic to this. Is the negation of
finitude incomprehensible? What would you think of a universe in which
you could travel one, ten, or a thousand miles, or even to California, and
then find it come to an end? Even if you were to go millions and millions
of miles, the idea of coming to an end is incomprehensible.

Popular Lectures and Addresses
Volume I
The Wave Theory of Light (pp. 314–5)

Tolstoy, Leo
And so to imagine the action of a man entirely subject to the law of
inevitability without any freedom, we must assume the knowledge of an
infinite number of space relations, an infinitely long period of time, and
an infinite series of causes.

War and Peace
Second Epilogue, Chapter X

Arriving at infinitesimals, mathematics, the most exact of sciences,
abandons the process of analysis and enters on the new process of the
integration of unknown, infinitely small, quantities.

War and Peace
Second Epilogue, Chapter XI

Unknown
Possibilities are infinite.

Source unknown

Infinity is where things happen that don't.

Quoted by W.W. Sawyer in
Prelude to Mathematics (p. 143)

Wolf, Fred Alan
Infinity is always one more than now.

Parallel Universes (p. 69)

Some scientists are also bothered by infinities, which seem to crop up at embarrassing places in our theories of the universe. A black hole . . . has an infinity at its very center.

Parallel Universes (p. 69)

I believe that the infinity of possibilities predicted to arise in quantum physics is the same infinity as the number of universe-possibilities predicted to arise in relativistic physics when, at the beginning of time, the universe, our home, and all of its sisters and brothers were created. As modest and troublesome as we often are, we too are never the less creatures of infinity.

Parallel Universes (p. 70)

Black holes are the bungholes of space.
Butthead – (See p. 34)

KNOWLEDGE

Aristotle
We think we have scientific knowledge when we know the cause, and there are four causes: (1) the definable form, (2) an antecedent which necessitates a consequent, (3) the efficient cause, (4) the final cause.

Posterior Analytics
Book II, Chapter 11, 94ᵃ, [20]

Yet it does not appear to be true in all cases that correlatives come into existence simultaneously. The object of knowledge would appear to exist before knowledge itself for it is usually the case that we acquire knowledge of objects already existing; it may be difficult, if not impossible, to find a branch of knowledge the beginning of the existence of which was contemporaneous with that of its object.

Categories
Chapter 7, 7ᵇ, [20]

Bacon, Francis
. . . that knowledge hath in it somewhat of the serpent, and therefore where it entereth into a man it makes him swell; *"scientia inflat* [knowledge puffs up]".

Advancement of Learning
First Book, Chapter I, 2

Billings, Josh
It is better to kno less, than to kno so mutch, that aint so.

Old Probability: Perhaps Rain—Perhaps Not

Borel, Émile
Incomplete knowledge must be considered as perfectly normal in probability theory; we might even say that, if we knew all the circumstances of a phenomenon, there would be no place for probability, and we would know the outcome with certainty.

Probability and Certainty
Chapter 1 (p. 13)

Byron, Lord George Gordon
That knowledge is not happiness, and science
But an exchange of ignorance for that
Which is another kind of ignorance.

The Poetical Works of Lord Byron
Manfred, a dramatic poem
Act 2, scene V, l. 431–3

Collingwood, Robin George
Questioning is the cutting edge of knowledge; assertion is the dead
weight behind the edge that gives it driving force.

Speculum Mentis (p. 78)

Cowper, William
Knowledge and Wisdom, far from being one,
Have oft-times no connexion. Knowledge dwells
In heads replete with thoughts of other men;
Wisdom in minds attentive to their own.
Knowledge, a rude unprofitable mass,
The mere materials with which wisdom builds . . .
Knowledge is proud that he has learn'd so much;
Wisdom is humble that he knows no more.

The Complete Poetical Works of William Cowper
The Task
Book VI, l. 88–93, 96, 97

da Vinci, Leonardo
The acquisition of any knowledge whatever is always useful to the
intellect, because it will be able to banish the useless things and retain
those which are good. For nothing can be either loved or hated unless it
is first known.

The Notebooks of Leonardo da Vinci
Volume I
Aphorisms (p. 88)

Dickens, Charles
But wot's that, you're a-doin' of? Pursuit of knowledge under difficulties,
Sammy?

Pickwick Papers
Chapter 33 (p. 456)

Einstein, Albert
Yet it is equally clear that knowledge of what *is*, does not open the door
directly to what *should be*.

Out of My Later Years (p. 22)

Fabing, Harold
Marr, Ray
Knowledge is a process of piling up facts; wisdom lies in their simplification.

Fischerisms (p. 2)

Holmes, Oliver Wendell
It is the province of knowledge to speak and it is the privilege of wisdom to listen.

The Poet at the Breakfast Table
Chapter 10

Scientific knowledge, even in the most modest persons, has mingled with it a something which partakes of insolence.

The Autocrat of the Breakfast-Table
Chapter 3

Huxley, Aldous
Knowledge is power and, by a seeming paradox, it is through their knowledge of what happens in this unexperienced world of abstractions and inferences that scientists and technologists have acquired their enormous and growing power to control, direct and modify the world of manifold appearances in which human beings are privileged and condemned to live.

Literature and Science
Chapter 3 (p. 9)

Huxley, Thomas
Indeed, if a little knowledge is dangerous, where is the man who has so much as to be out of danger?

Collected Essays
On Elementary Instruction in Physiology
Volume III

Jefferson, Thomas
A patient pursuit of facts, and cautious combination and comparison of them, is the drudgery to which man is subjected by his Master, if he wishes to attain sure knowledge.

Notes on the State of Virginia (p. 71, n)

Jevons, W.S.
I am convinced that it is impossible to expound the methods of induction in a sound manner, without resting them on the theory of probability. Perfect knowledge alone can give certainty, and in nature perfect knowledge would be infinite knowledge, which is clearly beyond

our capacities. We have, therefore, to content ourselves with partial knowledge,—knowledge mingled with ignorance, producing doubt.

The Principles of Science (p. 197)

Lichtenberg, Georg Christoph

I made the journey to knowledge like dogs who go for walks with their masters, a hundred times forward and backward over the same territory; and when I arrived I was tired.

Lichtenberg: Aphorisms & Letters
Aphorisms (p. 58)

Marlowe, Christopher

Our souls, whose faculties can comprehend
The wondrous architecture of the world,
And measure every wandering planet's course,
Still climbing after knowledge infinite . . .

Tamburlaine the Great
Part the First
Act II, scene 7, l. 20–3

Myrdal, Gunnar

All ignorance, like all knowledge, tends thus to be opportunist.

Objectivity in Social Research
Chapter III (p. 19)

Petit, Jean-Pierre

I'VE UNDERSTOOD IT! Well, that is . . . I'm not exactly sure WHAT I've understood, but I have the impression I've understood SOMETHING.

Euclid Rules OK? (p. 44)

Stewart, Ian

I may not understand it, but it sure looks important to me.

Does God Play Dice? (p. 121)

Szent-Györgyi, Albert

Knowledge is a sacred cow, and my problem will be how we can milk her while keeping clear of her horns.

Science
Teaching and the Expanding Knowledge (p. 1278)
Volume 146, Number 3649, 4 December 1964

Tennyson, Alfred Lord

Knowledge comes, but wisdom lingers.

The Poems and Plays of Tennyson
Locksley Hall
l. 141

LAWS

Babbage, Charles
The more man inquires into the laws which regulate the material universe, the more he is convinced that all its varied forms arise from the action of a few simple principles. These principles themselves converge, with accelerating force, towards some still more comprehensive law to which all matter seems to be submitted. Simple as that law may possibly be, it must be remembered that it is only one amongst an infinite number of simple laws: that each of these laws has consequences at least as extensive as the existing one, and therefore that the Creator who selected the present law must have foreseen the consequences of all other laws.

> Quoted by John D. Barrow in
> *Theories of Everything* (p. 16)

Bartusiak, Marcia
According to "Turner's Law," the invocation of the tooth fairy should not occur more than once in any scientific argument.

> *Thursday's Universe* (p. 207)

Dewey, John
Scientific principles and laws do not lie on the surface of nature. They are hidden, and must be wrested from nature by an active and elaborate technique of inquiry.

> *Reconstruction in Philosophy* (p. 32)

Euclid
The laws of nature are but the mathematical thoughts of God.

> Quoted by Stanley Gudder in
> *A Mathematical Journey* (p. 112)

Feynman, Richard P.
From a long view of the history of mankind—seen from, say, ten thousand years from now—there can be little doubt that the most

significant event of the 19th century will be judged as Maxwell's discovery of the laws of electrodynamics. The American Civil War will pale into provincial insignificance in comparison with this important scientific event of the same decade.

The Feynman Lectures on Physics
Volume II
Chapter 1-6 (p. 1-11)

There is also a rhythm and a pattern between the phenomena of nature which is not apparent to the eye, but only to the eye of analysis; and it is these rhythms and patterns which we call Physical Laws.

The Character of Physical Law
Chapter 1 (p. 13)

Freund, Peter
Think, for a moment, of a cheetah, a sleek, beautiful animal, one of the fastest on earth, which roams freely on the savannas of Africa. In its natural habitat, it is a magnificent animal, almost a work of art, unsurpassed in speed or grace by any other animal. Now, think of a cheetah that has been captured and thrown into a miserable cage in a zoo. It has lost its original grace and beauty, and is put on display for our amusement. We see only the broken spirit of the cheetah in the cage, not its original power and elegance. The cheetah can be compared to the laws of physics, which are beautiful in their natural setting. The natural habitat of the laws of physics is higher-dimensional space–time. However, we can only measure the laws of physics when they have been broken and placed on display in a cage, which is our three-dimensional laboratory. We only see the cheetah when its grace and beauty have been stripped away.

Quoted by Michio Kaku in
Hyperspace (p. 12)

Gamow, George
If and when all the laws governing physical phenomena are finally discovered, and all the empirical constants occurring in these laws are finally expressed through the four independent basic constants, we will be able to say that physical science has reached its end, that no excitement is left in further explorations, and that all that remains to a physicist is either tedious work on minor details or the self-educational study and adoration of the magnificence of the completed system. At that stage physical science will enter from the epoch of Columbus and Magellan into the epoch of National Geographic Magazine!

Physics Today
Any Physics Tomorrow? (p. 18)
Volume 2, Number 1, January 1949

Hill, Aaron
O'er Nature's laws God cast the veil of night,
Out-blaz'd a Newton's soul—and all was light.

On Sir Isaac Newton

Hutten, Ernest H.
In quantum physics, a law is a statement from which descriptions as well
as predictions of probability distribution of future events can be derived,
given certain conditions . . . a law is merely a statement, or theorem
within a scientific system.

The Language of Modern Physics (p. 222)

Huxley, Thomas
The chess-board is the world; the pieces are the phenomena of the
universe; the rules of the game are what we call the laws of Nature.
The player on the other side is hidden from us. We know that his play is
always fair, just, and patient. But also we know, to our cost, that he never
overlooks a mistake, or makes the smallest allowance for ignorance.

Lay Sermons, Addresses, and Reviews
A Liberal Education (pp. 31–2)

Koestler, A.
One branch after another of chemistry, physics and cosmology has
merged in the majestic river as it approaches the estuary—to be
swallowed up by the ocean, lose its identity, and evaporate into the
clouds; the final act of the great vanishing process, and the beginning,
one hopes, of a new cycle . . . .. It seems that the more universal the "laws"
which we discover, the more elusive they become, and that the ultimate
consummation of all rivers of knowledge is in the cloud of unknowing.

The Act of Creation (p. 252)

Laplace, Pierre Simon
All the effects of nature are only the mathematical consequences of a
small number of immutable laws.

Quoted by E.T. Bell in
Men of Mathematics (p. 172)

Pagels, Heinz R.
The fact that the universe is governed by simple natural laws is
remarkable, profound and on the face of it absurd. How can the vast
variety in nature, the multitude of things and processes all be subject to
a few simple, universal laws?

Perfect Symmetry (p. 160)

Instead of finding an absolute universal law at the bottom of existence, they may find an endless regress of laws, or even worse, total confusion and lawlessness—an outlaw universe.

Perfect Symmetry (p. 264)

Pearson, Karl
Scientific Law is description, not a prescription.

Quoted by Henry Crew in
General Physics (p. 45)

Planck, Max
Thus from the outset we can be quite clear about one very important fact, namely, that the validity of the law of causation for the world of reality is a question that cannot be decided on grounds of abstract reasoning.

Where is Science Going? (p. 113)

How do we discover the individual laws of Physics, and what is their nature? It should be remarked, to begin with, that we have no right to assume that any physical law exists, or if they have existed up to now, that they will continue to exist in a similar manner in the future. It is perfectly conceivable that one fine day Nature should cause an unexpected event to occur which would baffle us all; and if this were to happen we would be powerless to make any objection, even if the result would be that, in spite of our endeavors, we should fail to introduce order into the resulting confusion. In such an event, the only course open to science would be to declare itself bankrupt. For this reason, science is compelled to begin by the general assumption that a general rule of law dominates throughout Nature . . .

Universe in the Light of Modern Physics (pp. 58–9)

Russell, Bertrand
Scientific laws, when we have reason to think them accurate, are different in form from the common-sense rules which have exceptions: they are always, at least in physics, either differential equations, or statistical averages.

The Analysis of Matter
Chapter XIX (p. 191)

Schrödinger, Erwin
The laws of physics are generally looked upon as a paradigm of exactitude.

Science Theory and Man
Chapter VI (p. 133)

Snow, C.P.
Once or twice I have been provoked and have asked the company how many of them could describe the Second Law of Thermodynamics. The response was cold: it was also negative. Yet I was asking something which is about the scientific equivalent of: *Have you read a work of Shakespeare's?*

The Two Cultures (p. 16)

Squire, Sir John C.
Nature and Nature's laws lay hid in night:
God said, "Let Newton be!" and all was light.
But not for long. The devil howling, "Ho!
Let Einstein be!" restored the *status quo.*

Quoted by Cecilia Helena Payne Gasposchkin in
Introduction to Astronomy (p. 168)

Unknown
Ignorance of the law is no excuse.

Source unknown

Little laws have bigger laws
From which they're forced to follow,
But a bigger law's a bigger guess
That's harder still to swallow.

Quoted by John D. Barrow in
The World within the World (p. 236)

The Cartoon Laws of Physics

Cartoon Law I

Any body suspended in space will remain in space until made aware of its situation. Daffy Duck steps off a cliff, expecting further pastureland. He loiters in midair, soliloquizing flippantly, until he chances to look down. At this point, the familiar principle of 32 feet per second per second takes over.

Cartoon Law II

Any body in motion will tend to remain in motion until solid matter intervenes suddenly. Whether shot from a cannon or in hot pursuit on foot, cartoon characters are so absolute in their momentum that only a telephone pole or an oversized boulder retards their forward motion absolutely. Sir Isaac Newton called this sudden termination of motion the stooge's surcease.

Cartoon Law III

Any body passing through solid matter will leave a perforation conforming to its perimeter. Also called the silhouette of passage, this phenomenon is the specialty of victims of directed-pressure explosions and of reckless cowards who are so eager to escape that they exit directly through the wall of a house, leaving a cookie-cutout-perfect hole. The threat of skunks or matrimony often catalyses this reaction.

Cartoon Law IV

The time required for an object to fall twenty stories is greater than or equal to the time it takes for whoever knocked it off the ledge to spiral down twenty flights to attempt to capture it unbroken. Such an object is inevitably priceless, the attempt to capture it inevitably unsuccessful.

Cartoon Law V

All principles of gravity are negated by fear. Psychic forces are sufficient in most bodies for a shock to propel them directly away from the earth's surface. A spooky noise or an adversary's signature sound will induce motion upward, usually to the cradle of a chandelier, a treetop, or the crest of a flagpole. The feet of a character who is running or the wheels of a speeding auto need never touch the ground, especially when in flight.

Cartoon Law VI

As speed increases, objects can be in several places at once. This is particularly true of tooth-and-claw fights, in which a character's head may be glimpsed emerging from the cloud of altercation at several places simultaneously. This effect is common as well among bodies that are spinning or being throttled. A wacky character has the option of self-replication only at manic high speeds and may ricochet off walls to achieve the velocity required.

Cartoon Law VII

Certain bodies can pass through solid walls painted to resemble tunnel entrances; others cannot. This trompe l'oeil inconsistency has baffled generations, but at least it is known that whoever paints an entrance on a wall's surface to trick an opponent will be unable to pursue him into this theoretical space. The painter is flattened against the wall when he attempts to follow into the painting. This is ultimately a problem of art, not of science.

Cartoon Law VIII

Any violent rearrangement of feline matter is impermanent. Cartoon cats possess even more deaths than the traditional nine lives might comfortably afford. They can be decimated, spliced, splayed, accordion-pleated, spindled, or disassembled, but they cannot be destroyed. After a few moments of blinking self pity, they reinflate, elongate, snap back, or solidify.

Corollary: A cat will assume the shape of its container.

Cartoon Law IX

Everything falls faster than an anvil.

Source unknown

Voltaire
In effect, it would be very singular that all nature, all the planets, should obey eternal laws, and that there should be a little animal five feet high, who, in contempt of these laws, could act as he pleased, solely according to his caprice.

The Best Known Works of Voltaire
Ignorant Philosophers
Chapter XIII

Wheeler, John A.
There is no law except that there is no law.

Quoted by John D. Barrow in
The World within the World (p. 293)

Not only particles and the fields of force had to come into being at the big bang, but the laws of physics themselves, and this by a process as higgledy-piggledy as genetic mutation or the second law of thermodynamics.

International Journal of Theoretical Physics
The Computer and the Universe (p. 565)
Volume 21, Numbers 6/7, June 1982

Whitehead, Alfred North
The laws of physics are the decrees of fate.

Science and the Modern World
Chapter I (p. 16)

Wigner, Eugene
The miracle of the appropriateness of the language of mathematics for the formulation of the laws of physics is a wonderful gift which we neither understand nor deserve.

Symmetries and Reflections
The Unreasonable Effectiveness of Mathematics in the Natural Sciences (p. 237)

LIGHT

Alighieri, Dante
O Supreme Light, that so high upliftest Thyself from mortal conceptions, re-lend to my mind a little of what Thou didst appear, and make my tongue so powerful that it may be able to leave one single spark of Thy glory for the folk to come . . .

The Divine Comedy of Dante Alighieri
Paradise
Canto XXXIII, l. 67–72

Bohm, David
When we come to light, we are coming to the fundamental activity in which existence has its ground . . . .. Light is the potential of everything.

Quoted by Renée Weber in
Dialogues with Scientists and Sages (p. 155)

This ocean of energy could be thought of as an ocean of light.

Quoted by Renée Weber in
Dialogues with Scientists and Sages (p. 155)

Bragg, Sir William
Light brings us the news of the Universe.

The Universe of Light (p. 1)

de Fontenelle, Bernard
Light is composed of globules which rebound from a solid substance, but pass through any thing in which they find interstices, such as air or glass: the moon, therefore, gives us light in consequence of being a hard, solid body, which sends back these globules.

Conversations on the Plurality of Worlds
Second Evening (p. 36)

Dick, Thomas
Light is that invisible ethereal matter which renders objects perceptible by the visual organs. It appears to be distributed throughout the immensity of the universe, and is essentially requite to the enjoyment of every rank of perceptive existence. It is by the agency of this mysterious substance that we become acquainted with the beauties and sublimities of the universe, and the wonderful operations of the Almighty Creator.

The Works of Thomas Dick, LL.D.
Volume IX
The Practical Astronomer
Part I
On Light
Introduction (p. 191)

Eddington, Sir Arthur Stanley
Oh, leave the Wise our measures to collate.
One thing at least is certain, light has weight;
One thing is certain and the rest debate—
Light rays, when near the Sun, do not go straight.

Quoted by Peter Coveney and Roger Highfield in
The Arrow of Time (p. 93)

Einstein, Albert
All these fifty years of conscious brooding have brought me no nearer to the answer to the question "What are light quanta?" Nowadays every Tom, Dick, and Harry thinks he knows it, but he is mistaken.

Einstein: A Centenary Volume
Letter to M. Besso, 1951 (p. 138)

Feynman, Richard P.
I want to emphasize that light comes in this form—particles. It is very important to know that light behaves like particles, especially for those of you who have gone to school, where you were probably told something about light behaving like waves. I'm telling you the way it *does* behave—like particles.

QED (p. 15)

It isn't that a particle takes the path of least action but that it smells all the paths in the neighborhood and chooses the one that has the least action.

The Feynman Lectures on Physics
Volume II, Chapter 19 (p 19-9)

Genesis 1:3
Let there be light.

The Bible

Hoyle, Fred
Whales can make progress through water either by wagging their tails up and down or from side to side. Light can travel through space in two ways, one like a tail moving up and down and the other like the tail moving from side to side.

Frontiers of Astronomy (p. 253)

Jeans, Sir James Hopwood
If annihilation of matter occurs, the process is merely that of unbottling imprisoned wave-energy and setting it free to travel through space. These concepts reduce the whole universe to a world of light, potential or existent, so that the whole story of its creation can be told with perfect accurance and completeness in the six words: "God said, 'Let there be light.' "

The Mysterious Universe
Matter and Radiation (pp. 77–8)

La Science Populaire Editor
Light crosses space with the prodigious velocity of 6,000 leagues per second.

La Science Populaire
April 28, 1881

A typographical error slipped into our last issue that is important to correct. The speed of light is 76,000 leagues per hour—and not 6,000.

La Science Populaire
May 19, 1881

A note correcting a first error appeared in our issue number 68, indicating that the speed of light is 76,000 leagues per hour. Our readers have corrected this new error. The speed of light is approximately 76,000 leagues per second.

La Science Populaire
June 16, 1881

Maxwell, James Clerk
I have a paper afloat, with an electromagnetic theory of light, which 'til I am convinced to the contrary, I hold to be great guns.

Quoted by John N. Shive and Robert L. Weber in
Similarities in Physics (p. 123)

Miner, Virginia Scott
All color is to light as pitch to sound.
The Human eye can see one octave's light,
But those that soar past violet abound—
And octaves still exist, though not to sight,
Below the red.

The Physics Teacher
Physics inspires the Muses
Light (p. 635)
Volume 16, Number 9, December 1978

Newton, Sir Isaac
Are not the Rays of Light very small Bodies emitted from shining Substances?

Opticks
Book III, Part I
Query 29

Planck, Max
The velocity of light is to the Theory of Relativity as the elementary quantum of action is to the Quantum Theory: it is its absolute core.

Scientific Autobiography (p. 47)

Seneca
The red of the Dog star is brighter, that of Mars weaker; while Jupiter has no red, with its gleam extended into pure light.

Naturales Questiones
Volume III (p. 19)

Sommerfeld, Arnold
The twofold nature of light as a light-wave and as a light-quantum is thus extended to electrons and, further, to atoms: their wave-nature is asserting itself more and more, theoretically and experimentally, as concurrent with their corpuscular nature.

Wave Mechanics (p. 7)

Standen, Anthony
The velocity of light occupies an extraordinary place in modern physics. It is *lèse majesté* to make any criticism of the velocity of light. It is a sacred cow within a sacred cow, and it is just about the Absolutest Absolute in the history of human thought.

Science is a Sacred Cow (p. 73)

Unknown
How distant some of the nocturnal suns!
So distant, says the sage, 'twere not absurd
To doubt that beams set out at Nature's birth
Had yet arrived at this so foreign world,
Though nothing half so rapid as their flight!

Quoted by John Nichol in
Views of the Architecture of the Heavens

And God said:

$$\nabla \cdot E = \rho/\varepsilon_0$$
$$\nabla \cdot \beta = 0$$
$$\nabla \times E = -\partial\beta/\partial t$$
$$c^2\nabla \times \beta = j/\varepsilon_0 + \partial E/\partial t$$

And there was light.

Source unknown

Young, Joshua
There was once a sailor named Lee
Whose speed was much faster than "c".
But while racing his craft,
His bow followed his aft,
With a finish that no one could see.

The Physics Teacher
Physics Poems (p. 587)
Volume 20, Number 9, December 1982

Zee, Anthony
Let there be an SU(5) Yang–Mills theory with all its gauge bosons, let
the symmetry be broken down spontaneously, and let all but one of the
remaining massless gauge bosons be sold into infrared slavery. That one
last gauge boson is my favorite. Let him rush forth to illuminate all of
my creations!

Fearful Symmetry (p. 232)

LOCATION

Eddington, Sir Arthur Stanley

We have certain preconceived ideas about location in space which have come down to us from ape-like ancestors.

The Nature of the Physical World
Chapter I
"Commonsense" Objections (p. 16)

MAGNETIC

Blavatsky, H.P.
The earth is a magnetic body, in fact, as some scientists have found it is one vast magnet, as Paracelsus affirmed some 300 years ago.

Isis Unveiled
I (p. xxiii)

Dryden, John
Gilbert shall live till loadstones cease to draw
Or British fleets the boundless ocean awe.

The Poetical Works of Dryden
Epistle to Doctor Walter

Gilbert W.S.
Sullivan, Arthur
A magnet hung in a hardware shop,
And all around was a loving crop
Of scissors and needles, nails and knives,
Offering love for all their lives;
But for iron the magnet felt no whim,
Though he charmed iron, it charmed not him;
From needles and nails and knives he'd turn,
For he'd set his love on a Silver Churn!

While this magnetic,
Peripatetic
Lover he lived to learn,
By no endeavor
Can magnet ever
Attract a Silver Churn!

Patience
Act II

Gilbert, William

In like manner, the loadstone has from nature its two poles, a northern and a southern; fixed, definite points in the stone, which are the primary termini of the movements and effects, and the limits and regulators of the several actions and properties . . . whether its shape is due to design or chance, and whether it be long, or flat, or four-square, or three cornered, or polished; whether it be rough, broken-off, or unpolished: the loadstone ever has and ever shows its poles.

On the Loadstone and Magnetic Bodies
Book I, Chapter 3

Lucretius

Next in order I will proceed to discuss by what law of nature it comes to pass that iron can be attracted by that stone which the Greeks call the Magnet from the name of its native place, because it has its origin within the bounds of the country of the Magnesians.

On the Nature of Things
Book VI, Section 906

Parker, E.N.

It appears that the radical element responsible for the continuing thread of cosmic unrest is the magnetic field.

Cosmical Magnetic Fields (p. 2)

Magnetic fields (and their inevitable offspring fast particles) are found everywhere in the universe where we have the means to look for them.

Cosmical Magnetic Fields (p. 6)

Unknown

We know that the magnet loves the loadstone, but we do not know whether the loadstone also loves the magnet or is attracted to it against its will.

Inventing the Future (p. 217)

MATHEMATICS

Bacon, Francis
As Physic advances farther and farther every day and develops new axioms, it will require fresh assistance from Mathematic.

The Advancement of Learning
Book III, Chapter VI

Carlyle, Thomas
It is a mathematical fact that the casting of this pebble from my hand alters the centre of gravity of the universe.

Sartor Resartus
III

Courant, Richard
Since the seventeenth century, physical intuition has served as a vital source for mathematical problems and methods. Recent trends and fashions have, however, weakened the connection between mathematics and physics; mathematicians, turning away from the roots of mathematics in intuition, have concentrated on refinement and emphasized the postulated side of mathematics, and at times have overlooked the unity of their science with physics and other fields. In many cases, physicists have ceased to appreciate the attitudes of mathematicians. This rift is unquestionably a serious threat to science as a whole; the broad stream of scientific development may split into smaller and smaller rivulets and dry out. It seems therefore important to direct our efforts toward reuniting divergent trends by classifying the common features and interconnections of many distinct and diverse scientific facts.

Methods of Mathematical Physics (pp. v–vi)

D'Abro, A.
Success has attended the efforts of mathematical physicists in so large a number of cases that, however marvelous it may appear, we can scarcely

escape the conclusion that nature must be rational and susceptible to mathematical law.

The Evolution of Scientific Thought from Newton to Einstein (p. xii)

Dirac, Paul Adrien Maurice
From now on there will be no physical treatise which is not primarily mathematical.

Quoted by Walter R. Fuchs in
Mathematics for the Modern Mind (p. 25)

The steady progress of physics requires for its theoretical formulations a mathematics that gets continually more advanced.

Proceedings of the Royal Society
Quantisized Singularities in the Electromagnetic Field (p. 60)
Series A, Volume 133, Number A821, 1931

Dyson, Freeman J.
On being asked what he meant by the beauty of a mathematical theory of physics, Dirac replied that if the questioner was a mathematician then he did not need to be told, but were he not a mathematician then nothing would be able to convince him of it.

Quoted by John D. Barrow in
Theories of Everything (p. 16)

Einstein, Albert
Experience remains, of course, the sole criterion of the physical utility of a mathematical construction.

The World As I See It (p. 36)

Our experience hitherto justifies us in believing that nature is the realisation of the simplest conceivable mathematical ideas. I am convinced that we can discover by purely mathematical constructions the concepts and the laws connecting them with each other, which furnish the key to the understanding of natural phenomena. Experience may suggest the appropriate mathematical concepts, but they most certainly cannot be deduced from it. Experience remains, of course, the sole criterion of the utility of a mathematical construction. But the creative principle resides in mathematics. In a certain sense, therefore, I hold it true that pure thought can grasp reality as the ancients dreamed.

The World As I See It (pp. 36–7)

Thus the partial differential equation entered theoretical physics as a handmaid, but has gradually become mistress.

The World As I See It (p. 63)

At this point an enigma presents itself which in all ages has agitated inquiring minds. How can it be that mathematics, being after all a product of human thought which is independent of experience, is so admirably appropriate to the objects of reality?

Sidelights on Relativity (p. 28)

One reason why mathematics enjoys special esteem, above all other sciences, is that its laws are absolutely certain and indisputable, while those of all other sciences are to some extent debatable and in constant danger of being overthrown by newly discovered facts.

Sidelights on Relativity (p. 27)

Physics . . . is essentially an intuitive and concrete science. Mathematics is only a means for expressing the laws that govern phenomena.

Quoted by Maurice Solovine in
Einstein: A Centenary Volume
Excerpts from a memoir (p. 9)

Don't worry about your difficulties in mathematics; I can assure you that mine are still greater.

Source unknown

Farrar, John
. . . in mathematical science, and in it alone, man sees things precisely as God sees them, handles the very scale and compass with which the Creator planned and built the universe . . .

As reported by Andrew P. Peabody in Florian Cajori's
The Teaching and History of Mathematics in the United States (fn, p. 128)

Feynman, Richard P.
Now you may ask, "What is mathematics doing in a physics lecture?" We have several possible excuses: first, of course, mathematics is an important tool, but that would only excuse us for giving the formula in two minutes. On the other hand, in theoretical physics we discover that all our laws can be written in mathematical form; and that this has a certain simplicity and beauty about it. So, ultimately, in order to understand nature it may be necessary to have a deeper understanding of mathematical relationships. But the real reason is that the subject is enjoyable, and although we humans cut nature up in different ways, and we have different courses in different departments, such compartmentalization is really artificial, and we should take our intellectual pleasures where we find them.

The Feynman Lectures on Physics
Volume 1
Chapter 22-1 (p. 22-1)

Gibbs, J. Willard
Maxwell's Treatise on Electricity and Magnetism has done so much to familiarize students of physics with quaternion notations that it seems impossible that this subject should ever again be entirely divorced from the methods of multiple algebra.

I wish that I could say as much of astronomy. It is, I think, to be regretted that the oldest of the scientific applications of mathematics, the most dignified, the most conservative, should keep so far aloof from the youngest of mathematical methods.

Quoted by Florian Cajori in
The Teaching and History of Mathematics in the United States (p. 157)

Jeans, Sir James Hopwood
The essential fact is simply that *all* the pictures which science now draws of nature, and which alone seem capable of according with observational fact, are mathematical pictures.

The Mysterious Universe
Into Deep Waters (p. 127)

Jeffreys, Sir Harold
Jeffreys, B.S.
We have to come back to something like ordinary language after all when we want to talk *about* mathematics!

Methods of Mathematical Physics (p. 2)

Lillich, Robert
Divergence B, it's plain to see, is zero.
  I think they've got it, I think they've got it!
And del dot D is always rho, you know.
  I think they've got it, I think they've got it!

The Physics Teacher
My Fair Physicist (p. 490)
I Think They've Got It
Volume 6, Number 9, December 1968

Montague, W.P.
. . . that climax of rationalists' aspiration in which even the arbitrary and existential constants of physics would be reduced to a crystalline precipitate of purely subsistential mathematical relations.

Philosophical Review
The Einstein Theory and the Possible Alternative (p. 169)
Volume 33, Number 2, March 1924

Roentgen, Wilhelm Conrad

The physicist in preparing for his work needs three things, mathematics, mathematics, and mathematics.

Quoted by E.B. Escott
The Mathematical Gazette
Gleanings Far and Near
Volume 22, Number 252, December 1938
Number 1225

Unknown

I'm very well acquainted too with matters mathematical,
I understand equations, both ordinary and differential,
About Riemann's geometry I'm teeming with a lot o'news—
With many new facts about the square of the hypotenuse.

I'm very good at black holes, be they charged or spinning,
About singularities I know more than Penrose and Hawking;
In short, in matters vegetable, animal and mineral,
I am the very model of a modern Major General.

Quoted by Jayant Narlikar in
Violent Phenomena in the Universe (p. 52)

von Neumann, John

Mathematics falls into a great number of subdivisions, differing from one another widely in character, style, aims, and influence. It shows the very opposite of the extreme concentration of theoretical physics. A good theoretical physicist may today still have a working knowledge of more than half of his subject. I doubt that any mathematician now living has much of a relationship to more than a quarter.

Quoted by Cecil M. DeWitt and John A. Wheeler in
Battelle Rencontres (p. ix)
1967 Lectures in Mathematics and Physics

Whewell, William

The persons who have been employed on these problems of applying the properties of matter and the laws of motion to the explanation of the phenomena of the world, and who have brought to them the high and admirable qualities which such an office requires, have justly excited in a very eminent degree the admiration which mankind feels for great intellectual powers. Their names occupy a distinguished place in literary history; and probably there are no scientific reputations of the last century higher, and none more merited, than those earned by great mathematicians who have laboured with such wonderful success in unfolding the mechanisms of the heavens; such for instance as D'Alembert, Clairaut, Euler, Lagrange, Laplace.

Astronomy and General Physics
Book 3, Chapter 4 (p. 327)

Zee, Anthony

Toward the end of the last century, many physicists felt that the mathematical description of physics was getting ever more complicated. Instead, the mathematics involved has become ever more abstract, rather than more complicated. The mind of God appears to be abstract but not complicated. He also appears to like group theory.

Fearful Symmetry
Chapter 9 (p. 132)

I THINK WE MUST HAVE REPRODUCED....

Because physicists are a small group . . . they tend to behave like an amoeba.

Howard A. Robinson – (See p. 202)

MATTER

Bohm, David
Matter is like a small ripple on this tremendous ocean of energy, having some relative stability and being manifest.

Quoted by Renée Weber in
Dialogues with Scientists and Sages (p. 51)

Boswell, James
I shall always remember the alacrity with which Johnson answered, striking his foot with mighty force against a large stone, till he rebounded from it—"I refute it thus".

Boswell's Life of Johnson
6 August, 1763

Buller, Arthur Henry Reginald I.
To her friends said the Bright one in chatter,
"I have learned something new about matter:
  As my speed was so great
  Much increased was my weight,
Yet I failed to become any fatter."

Quoted by Clifton Fadiman in
The Mathematical Magpie (p. 290)

Čapek, Milič
This concept [matter] has hardly changed from the times of Leucippus to the beginning of the twentieth century: an impenetrable *something*, which fills completely certain regions of space and which persists through time even when it changes its location.

The Philosophical Impact of Contemporary Physics (p. 54)

Chaudhuri, Harides
We know too much about matter today to be materialists any longer.

The Philosophy of Integralism (p. 146)

Dewey, John
It would be difficult to find a greater distance between any two terms than that which separates "matter" in the Greek–medieval tradition and the technical signification, suitably expressed in mathematical symbols, that the word bears in science today.

In Yervant H. Krikorian's (Editor)
Naturalism and the Human Spirit
Antinaturalism in extremis (p. 3)

Dyson, Freeman J.
We have learned that matter is weird stuff. It is weird enough, so that it does not limit God's freedom to make it do what he pleases.

Infinite in All Directions
Chapter 1 (p. 8)

Hawking, Stephen
The point is that the new raw material doesn't really have to come from anywhere . . .. The universe can start off with zero energy and still create matter.

Quoted by Renée Weber in
Dialogues with Scientists and Sages (p. 51)

Kant, Immanuel
Give me matter and I will construct a world out of it!

Kant's Cosmogony
Preface (p. 29)

Kline, Morris
. . . where is the good, old-fashioned, solid matter that obeys precise, compelling mathematical laws? The stone that Dr. Johnson once kicked to demonstrate the reality of matter has become dissipated in a diffuse distribution of mathematical probabilities.

Mathematics in Western Culture (p. 382)

Newton, Sir Isaac
The quantity of matter is the measure of the same, arising from its density and bulk conjointly.

Mathematical Principles of Natural Philosophy
Definitions, Definition I

. . . it seems probable to me that God in the beginning formed matter in solid, massy, hard, impenetrable, moveable particles, of such sizes and figures, and with such other properties, and in such proportion to space, as most conduced to the end for which he formed them; and that these primitive particles being solids, are incomparably harder than any porous bodies compounded of them; even so very hard as never to wear or break in pieces; no ordinary power being able to divide what God himself made one in the first creation.

Opticks
Book II, Part I (very near the end)

As to your first query, it seems to me that if the matter of our sun and planets and all the matter of the universe were evenly scattered throughout all the heavens, and every particle had an innate gravity toward all the rest, and the whole space throughout which this matter was scattered was but finite, the matter on the outside of this space would, by its gravity, tend toward all the matter on the inside and, by consequence, fall down into the middle of the whole space and there compose one great spherical mass.

Theories of the Universe
I
Four Letters to Richard Bentley (p. 211)
December 10, 1692

Poe, Edgar Allan
. . . matter exists only as attraction and repulsion—that attraction and repulsion are matter.

The Complete Works of Edgar Allan Poe
Eureka (p. 214)

Prigogine, Ilya
When matter is becoming disturbed by non-equilibrium conditions it organizes itself, it wakes up. It happens that our world is a non-equilibrium system.

Quoted by Renée Weber in
Dialogues with Scientists and Sages (p. 51)

Santayana, George
. . . and the animate world must needs concern the physicist, since it is the crown of nature, the focus where matter concentrates its fires and best shows what it is capable of doing.

The Realm of Matter (p. 141)

Shelley, Percy Bysshe
I change but I cannot die.

The Complete Poetical Works of Shelley
The Cloud
l. 76

Spencer, Theodore
Matter whose movement moves us all
Moves to its random funeral,
And Gresham's law that fits the purse
Seems to fit the universe.

Reproduced in Helen Plotz's
Imagination's Other Place
Entropy

Thomson, Sir J.J.
From the point of view of the physicist, a theory of matter is a policy rather than a creed; its object is to connect or co-ordinate apparently diverse phenomena, and above all to suggest, stimulate and direct experiment. It ought to furnish a compass which, if followed, will lead the observer further and further into previously unexplored regions.

The Corpuscular Theory of Matter
Chapter I (p. 1)

Thomson, William [Lord Kelvin]
If the motion of every particle of matter in the universe were precisely reversed at any instant, the course of nature would be simply reversed for ever after. The bursting bubble of foam at the foot of a waterfall would reunite and descend into the water; the thermal motions would reconcentrate their energy, and throw the mass up the fall in drops reforming in a close column of ascending water . . . living creatures would grow backwards, with conscious knowledge of the future, but no memory of the past, and would become again unborn.

Quoted by John D. Barrow in
The World within the World (p. 126)

Young, Edward
"Has Matter *innate* Motion? Then each Atom,
"Asserting its indisputable Right
"To dance, would form an Universe of Dust:
"Has Matter *none*? Then whence these glorious Forms,
"And boundless Flights, from *Shapeless*, and *Repos'd*?. . ."

Night Thoughts
Night IX, l. 1472–6

METEORS

Bryant, Edward
METEOR
  METEOR
    METEOR
      METEOr
        METEOr
          METEOr
            METEor
              METEor
                METEor
                  METeor
                    METeor
                      METeor
                        MEteor
                          MEteor
                            MEteor
                              Meteor
                                Meteor
                                  Meteor
                                    meteor
                                      meteorite

Fantasy and Science Fiction
Winslow Crater (p. 82)
Volume 56, Number 3, March 1979

Eddington, Sir Arthur Stanley
I would rather believe in ghosts than in hyperbolic meteors.

Quoted by David H. Levy in
The Man Who Sold the Milky Way (p. 8)

Jefferson, Thomas
I could more easily believe that two Yankee professors would lie than that stones would fall from the heaven.

In R.V. Jones'
Physics Bulletin
The Natural Philosophy of Flying Saucers (p. 225)
Volume 19, 1968

Joshua 10:11
... the Lord cast down great stones from heaven upon them unto Azekah, and they died.

The Bible

London, Jack
I would rather be a meteor, every atom of
    me in magnificent glow
    than a sleepy and permanent planet.

Quoted by *StarDate*
May/June 1955 (p. 3)

Martin, Florence Holcomb
Slashed by the earth the comet's orbit glares
With tiny meteors; each fiery tail
Now into incandescence sparks and flares
In earth's rare upper atmosphere.

The Scientific Monthly
The Riddle of the Skies (p. 119)
Volume LXXV, Number 2, August 1952

Twain, Mark
We used to watch the stars that fell, too, and see them streak down. Jim allowed they'd got spoiled and was hove out of the nest.

Huckleberry Finn
Chapter XX

Virgil
As oft, from heaven unfixed, shoot flying stars,
And trail their locks behind them.

The Aeneid
V, l. 528–9

MILKY WAY

Chaucer, Geoffrey
See younder, lo, the Galaxye
Which men clepeth the Milky Wey,
For hit is whyt . . .

The Hous of Fame
Book II

de Fontenelle, Bernard
. . . you see that white part of the sky, called the milky-way. Can you guess what it is?—An infinity of little stars, invisible to the eyes on account of their smallness, and placed so close to each other that they seem but a stream of light. I wish I had a telescope here to shew you this cluster of worlds.

Conversations on the Plurality of Worlds
Fifth Evening (pp. 117–8)

de Morgan, Augustus
I have often had the notion that all the nebulæ we see, including our own, which we call the Milky Way, may be particles of snuff in the box of a giant of a proportionately larger universe. Of course the minimum time—a million of years or whatever the geologists make it—which our little affair has lasted, is but a very small fraction of a second to the great creature in whose nose we shall all be in a few tens of thousands of millions of millions of millions of years.

A Budget of Paradoxes
Volume II (pp. 191–2)

Lambert, Heinrich
I am undecided whether or not the visible Milky Way is but one of countless others all of which form an entire system. Perhaps the light from these infinitely distant galaxies is so faint that we cannot see them.

Quoted by Eli Maor in
To Infinity and Beyond: A Cultural History of the Infinite (p. 217)

Longfellow, Henry Wadsworth
Showed the broad, white road in heaven,
Pathway of the ghosts, the shadows,
Running straight across the heavens,
Crowded with the ghosts, the shadows.

The Poetical Works of Henry Wadsworth Longfellow
The Song of Hiawatha
Hiawatha's Childhood

Milton, John
A broad and ample road, whose dust is Gold,
And pavement Starrs, as Starrs to thee appear,
Seen in the galaxie, that Milkie way . . .

Paradise Lost
Book VII, l. 577–8

Ovid
There is a way on high, conspicuous in the clear heavens, called the
Milky Way, brilliant with its own brightness.

Metamorphoses
Book 1, l. 168

Pasternak, Boris
And there, with frightful listing
Through emptiness, away
Through unknown solar systems
Revolves the Milky Way . . .

Boris Pasternak: Fifty Poems
Night

Poincaré, Henri
Consider now the Milky Way; there also we see an innumerable dust;
only the grains of this dust are not atoms, they are stars; these grains
move also with high velocities; they act at a distance one upon another,
but this action is so slight at great distance that their trajectories are
straight; and yet, from time to time, two of them may approach near
enough to be deviated from their path, like a comet which had passed
too near Jupiter. In a world, to the eyes of a giant for whom our suns
would be as for us our atoms, the Milky Way would seem only a bubble
of gas.

The Foundations of Science
Science and Method
The Milky Way and the Theory of Gases (p. 524)

Updike, John
The Milky Way, which used to be thought of as the path by which the souls of the dead traveled to Heaven, is an optical illusion; you could never reach it. Like fog, it would always thin out around you. It's a mist of stars we make by looking the long way through the galaxy . . .

The Centaur (p. 37)

At night astronomers agree . . .
Matthew Prior – (See p. 6)

MODELS

Box, G.E.P.
All models are wrong but some are useful.

In R.L. Launer and G.N. Wilkinson (Editors)
Robustness in Statistics
Robustness in the Strategy of Scientific Model Building

Bronowski, Jacob
The pre-eminence of astronomy rests on the peculiarity that it can be treated mathematically; and the progress of physics, and most recently biology, has hinged equally on finding formulations of their laws that can be displayed as mathematical models.

The Ascent of Man (p. 165)

Davies, Paul Charles William
The incorporation of imaginary elements into physical theories is one of the most difficult practices for a professional physicist to justify to the layman. Of course, if a particular feature, such as isotopic spin symmetry, renders the model a brilliant success, then the physicist can simply reply, "I put it in because it works!"

Superforce (pp. 66–7)

Eigen, Manfred
A theory has only the alternatives of being right or wrong. A model has a third possibility: it may be right, but irrelevant.

In Jagdish Mehra's
The Physicist's Conception of Nature (p. 618)

Hutten, Ernest H.
There are . . . no mathematical models in physics: the equation by itself is not the model.

The Language of Modern Physics (p. 290)

Jeans, Sir James Hopwood
The making of models . . . to explain mathematical formulae and the phenomena they describe, is not a step towards, but a step away from reality; it is like making graven images of a spirit . . .. All the same, the mathematical physicist is still busily at work making graven images of the concepts of the wave-mechanics.

The Mysterious Universe
Into the Deep Waters (p. 141)

Thomson, William [Lord Kelvin]
I never satisfy myself until I can make a mechanical model of a thing. If I can make a mechanical model, I understand it.

Baltimore Lectures on Molecular Dynamics, and the Wave Theory of Light (p. 270)

MONOPOLES

Dirac, Paul Adrien Maurice
From the theoretical point of view one would think that monopoles should exist, because of the prettiness of the mathematics. Many attempts to find them have been made, but all have been unsuccessful. One should conclude that pretty mathematics by itself is not an adequate reason for nature to have made use of a theory. We still have much to learn in seeking for the basic principles of nature.

Quoted by Heinz R. Pagels in
Perfect Symmetry (p. 284)

Gamow, George
Two Monopoles worshipped each other,
And all of their sentiments clicked.
Still, neither could get to his brother,
Dirac was so fearfully strict!

Thirty Years that Shook Physics (p. 202)

MOTION

Aristotle

. . . the downward movement of a mass of gold or lead, or of any other body endowed with weight, is quicker in proportion to its size.

On the Heavens
Book IV, 2

We see that bodies which have a greater impulse either of weight or of lightness, if they are alike in other respects, move faster over an equal space, and in the ratio which their magnitudes bear to each other.

Physics
Book IV, 8

Carroll, Lewis

"Well, in *our* country," said Alice, still panting a little, "you'd generally get to somewhere else—if you ran very fast for a long time, as we've been doing."

"A slow sort of country!" said the Queen. "Now here, you see, it takes all the running you can do, to keep in the same place. If you want to get somewhere else, you must run at least twice as fast as that."

The Complete Works of Lewis Carroll
Through the Looking Glass
The Garden of Live Flowers

Cohen, I. Bernard

Odd as it may seem, most people's views about motion are part of a system of physics that was proposed more than 2,000 years ago and was experimentally shown to be inadequate at least 1,400 years ago.

The Birth of New Physics (p. 3)

165

Dee, John
Whatever is in the universe is continuously moved by some species of motion.

John Dee On Astronomy
XVI

Galilei, Galileo
My purpose is to set forth a very new science dealing with a very ancient subject. There is, in nature, perhaps nothing older than motion, concerning which the books written by philosophers are neither very few nor small; nevertheless, I have discovered by experiment some properties of it which are worth knowing and which have not hitherto been either observed or demonstrated.

Dialogues Concerning Two New Sciences
Third Day
Change of Position

. . . accelerated motion remains to be considered.

And first of all it seems desirable to find and explain a definition best fitting natural phenomena. For anyone may invent an arbitrary type of motion and discuss its properties . . . but we have decided to consider the phenomena of bodies falling with an acceleration such as actually occurs in nature and to make this definition of accelerated motion exhibit the essential features of observed accelerated motion.

Dialogues Concerning Two New Sciences
Naturally Accelerated Motion

Huygens, Christiaan
It is inconceivable to doubt that light consists in the motion of some sort of matter: For when one considers its production, one sees that here upon the earth it is chiefly engendered by fire and flame which contain without doubt bodies that are in rapid motion, since they dissolve and melt many other bodies, even the most solid; or when one considers its effects, one sees that when light is collected, as by concave mirrors, it has the property of burning as a fire does, that is to say, it disunites the particles of bodies. This is assuredly the mark of motion, at least in the true philosophy, in which one conceives the cause of all natural effects in terms of mechanical motions. This, in my opinion, we must necessarily do, or else renounce all hopes of ever comprehending anything in physics.

Treatise on Light
Chapter One
On Rays Propagated in Straight Lines

Jeans, Sir James Hopwood
. . . the laws which nature obeys are less suggestive of those which a machine obeys in its motion than of those which a musician obeys

in writing a fugue, or a poet in composing a sonnet. The motions of electrons and atoms do not resemble those of the parts of a locomotive so much as those of the dancers in a cotillion. And if the "true essence of substances" is for ever unknowable, it does not matter whether the cotillion is danced at a ball in real life, or on a cinematography screen, or in a story of Boccaccio.

The Mysterious Universe
Into Deep Waters (p. 136)

Lucretius
For whenever bodies fall through water and thin air, they must quicken their descents in proportion to their weights, because the body of water and subtle nature of air cannot retard everything in equal degree, but more readily give way, overpowered by the heavier . . .

On the Nature of Things
Book 2, l. 230–4

Newton, Sir Isaac
The quantity of motion is the measure of the same, arising from the velocity and quantity of matter conjointly.

Mathematical Principles of Natural Philosophy
Definitions, Definition II

. . . from the phenomena of motions to investigate the forces of nature, and then from these forces to demonstrate the other phenomena . . . the motions of the planets, the comets, the moon and the sea . . .

Quoted by Eric M. Rogers in
Physics for the Inquiring Mind (p. 1)

Raman, V.V.
Once during mass, Galileo in church
Conducted a major scientific search.
He measured with his pulse how a
   lamp did swing
That was to the ceiling tied with a
   string.

The Physics Teacher
A Fable for Physicists
The Pendulum Period (p. 488)
Volume 18, Number 7, October 1990

Sarpi, Fra Paolo
To give us the science of motion God and Nature have joined hands and created the intellect of Galileo.

Quoted by Morris Kline in
Mathematics and the Physical World (p. 181)

Unknown
Brownian motion: girl scout jogging.

Source unknown

Walters, Marcia C.
The fact that the photon gets mass from its motion
Is a widely accepted Einsteinian notion,
This doesn't apply to we mortals, alas—
For the smaller our motion the greater our mass.

The Physics Teacher
Filler (p. 384)
Volume 5, Number 8, November 1967

Wells, H.G.
"And here," he said, and opened the hand that held the glass. Naturally I winced, expecting the glass to smash. But so far from smashing, it did not even seem to stir; it hung in mid-air—motionless. "Roughly speaking," said Gibberne, "an object in these latitudes falls 16 feet in the first second. This glass is falling 16 feet in a second now. Only, you see, it hasn't been falling yet for the hundredth part of a second. That gives you some idea of the pace of my Accelerator."

28 Science Fiction Stories of H.G. Wells
The New Accelerator

Whitman, Walt
The universe is a procession with measured and beautiful motion.

Complete Poetry and Collected Prose
Leaves of Grass

Wittgenstein, Ludwig
The fact that we can describe the motions of the world using Newtonian mechanics tells us nothing about the world. The fact that we do, does tell us something about the world.

Quoted by John D. Barrow in
The World within the World (p. 77)

Young, Joshua
Said the earth to a ball falling free,
"You're enjoying this falling, I see."
The ball widened its eyes
And remarked with surprise,
"But it's you who is falling, not me!"

The Physics Teacher
Physics Poems (p. 587)
Volume 20, Number 9, December 1982

NATURE

de Fontenelle, Bernard
. . . nature so entirely conceals from us the means by which her scenery is produced, that for a long time we were unable to discover the causes of her most simple movements.

Conversations on the Plurality of Worlds
First Evening (p. 9)

Montaigne, Michel Eyquen
Let us a little permit Nature to take her own way; she better understands her own affairs than we.

Essays
Experience

Planck, Max
If one wishes to obtain a definite answer from Nature one must attack the question from a more general and less selfish point of view.

A Survey of Physics (p. 15)

Shakespeare, William
In Nature's infinite book of secrecy
A little I can read.

Anthony and Cleopatra
Act I, scene 2, l. 9–10

NEUTRINO

Adams, Douglas
The chances of a neutrino actually hitting something as it travels through all this howling emptiness are roughly comparable to that of dropping a ball bearing at random from a cruising 747 and hitting, say, an egg sandwich.

Mostly Harmless
Chapter 3

Eddington, Sir Arthur Stanley
In an ordinary way I might say that I do not believe in neutrinos. But I have to reflect that a physicist may be an artist, and you never know where you are with artists. My old-fashioned kind of disbelief in neutrinos is scarcely enough. Dare I say that experimental physicists will not have sufficient ingenuity to *make* neutrinos? Whatever I may think, I am not going to be lured into a wager against the skill of the experimenters under the impression that it is a wager against the truth of a theory. If they succeed in making neutrinos, perhaps even in developing industrial application of them, I suppose I shall have to believe—though I may feel they have not been playing quite fair.

The Philosophy of Physical Science (p. 112)

Gamow, George
My mass is zero,
  My Charge is the same.
You are my hero,
  Neutrino's my name.

Thirty Years that Shook Physics (p. 188)

. . . one of the students asked whether the "Chadwick neutron" was the same "neutron" proposed by Pauli for the phenomena of beta transformation. "No," answered Fermi, "il neutrone di Pauli è molto più piccolo, cio è un neutrino." The name stuck . . .

Physics Today
The Reality of Neutrinos (p. 5)
Volume 1, Number 3, July 1948

Haag, Jole
The Poet, I, Alfred Neutrino
Who subsisted sublimely on vino,
  With a spin of one-half
  Wrote his own epitaph:
"No rest-mass, no charge, no bambino."

Quoted by R.L. Weber in
A Random Walk in Science (p. 138)

Harari, Haim
Neutrino physics is largely an art of learning a great deal by observing nothing.

Source unknown

Pauli, Wolfgang
I have committed the ultimate sin, I have predicted the existence of a particle that can never be observed.

Quoted by John D. Barrow and B. Devine in
Longing for the Harmonies (p. 65)

Perry, Georgette
To trap them is almost impossible.
You may wait for months in a deep mine
Inside an anti-coincidence shield.
No charge deflects them.
Desireless, they cruise through the world
As if it's nothing, not there.

Twigs
Neutrinos

Ruderman, M.A.
Rosenfeld, A.H.
Every second, hundreds of billions of these neutrinos pass through each square inch of our bodies, coming from above during the day and from below at night, when the sun is shining on the other side of the earth!

American Scientist
An Elementary Statement on Elementary Particle Physics (p. 214)
Volume 48, Number 2, June 1960

Unknown

The neutrino is about as close to intangibility as we can get in this world—the human soul, perhaps is the next stage.

Engineering and Science
February 1973 (p. 15)

Mister Jordan
Takes neutrinos
And from those he
Builds the light.
And in pairs they
Always travel
One neutrino's
Out of sight.

Quoted by Abraham Pais in
Inward Bound (p. 419)
to the tune "Mac the Knife"

Updike, John

Neutrinos, they are very small.
They have no charge and have no mass
And do not interact at all.

Telephone Poles and Other Poems
Cosmic Gall

NEUTRON

Chadwick, J.
I think we shall have to make a real search for the neutron.

Quoted by A.S. Eve in
Rutherford
Letter to E. Rutherford (p. 300)

Gamow, George
The *Neutron* has come to be.
Loaded with Mass is he.
Of Charge, forever free.
Pauli, do you agree?

Thirty Years that Shook Physics (p. 214)

Heisenberg, Werner
The basic idea is: shove all fundamental difficulties on to the neutron and practice quantum mechanics inside the nucleus.

Quoted by Abraham Pais in
Inward Bound
Letter to Niels Bohr, 20 June 1932 (p. 413)

Unknown
Pity the poor neutron,
He thought he was a proton,
But he wasn't positive.

The Physics Teacher
Filler (p. 301)
Volume 22, Number 5, May 1984

NEWTON, SIR ISAAC

Blake, William
. . . May God us keep
From Single vision & Newton's Sleep!

The Letters of William Blake
Letter to Thomas Butt (p. 79)
Letter #30, November 22, 1802

Einstein, Albert
No one must think that Newton's great creation can be overthrown in any real sense by this or any other theory. His clear and wide ideas will forever retain their significance as the foundation on which our modern conceptions of physics have been built.

Out of My Later Years
Time, Space and Gravitation (p. 58)

Emerson, Ralph Waldo
Newton and Laplace need myriad's of ages and thick-strewn celestial areas. One may say a gravitating solar system is already prophesied in the nature of Newton's Work.

The Works of Ralph Waldo Emerson
Volume II
History

Gilman, Greer
Once upon a time in the $(n-1)$th dynasty of the Kingdom of Pendleton, there lived a high & mighty lord who was called Sir Isaac Fig Newton.

The Physics Teacher
Ergo (p. 514)
Volume 8, Number 9, December 1970

Hill, Thomas
The discoveries of Newton have done more for England and for the race, than has been done by whole dynasties of British monarchs.

Quoted by David M. Burton in
The History of Mathematics (p. 325)

Lee, A.
GRAVITY, GRAVITY,
Keeps us on the ground
An apple fell on Newton,
He said, "What goes up comes down."

The Physics Teacher
What is Christmas without (Physics) Carols?
Gravity (p. 449)
Volume 22, Number 7, October 1984

Newton's Epitaph
Who, by a vigor of mind almost divine, the motions and figures of the planets, the paths of comets, and the tides of the seas first demonstrated.

Quoted by Morris Kline in
Mathematical Thought from Ancient to Modern Times (p. 342)

Noyes, Alfred
Could Rembrandt but have painted him,
    in those hours
Making his first analysis of light
Alone, there, in his darkened Cambridge
    room
At Trinity! . . .

Watchers of the Sky
Newton, I

Pope, Alexander
Nature and Nature's laws lay hid in Night.
God said: "*Let* Newton *be*"; and all was Light.

The Complete Poetical Works of POPE
Epitaph intended for Sir Isaac Newton

Russell, Bertrand
This latter objection was sanctioned by Newton, who was not a strict Newtonian.

The Analysis of Matter
Chapter II (p. 14)

Thomson, James
Even light itself, which every thing displays,
Shone undiscovered, till his brighter mind
Untwisted all the shining robe of day;
And, from the whitening undistinguished blaze,
Collecting every ray into his kind,
To the charmed eye educed the gorgeous train
Of parent colours . . .

The Complete Poetical Works of James Thomson
To the Memory of Newton
l. 96–102

Whitehead, Alfred North
His cosmology [Newton's] is very easy to understand and very hard to believe.

Adventures of Ideas (p. 168)

. . . May God us keep
From Single vision & Newton's Sleep!
William Blake – (See p. 174)

OBSERVATION

Anscombe, F.J.
No observations are absolutely trustworthy.

Technometrics
Rejection of Outliers (p. 124)
Volume 2, Number 2, May 1960

Aristotle
. . . while those whom devotion to abstract discussions has rendered unobservant of the facts are too ready to dogmatize on the basis of a few observations.

On Generation and Corruption
Book I, Chapter II

Aurelius, Marcus [Antoninus]
Consider that everything which happens, happens justly, and if thou observest carefully, thou wilt find it to be so.

The Meditations of Marcus Aurelius
Book IV, Section 10

Bernard, Claude
Speaking concretely, when we say "making experiments or making observations," we mean that we devote ourselves to investigation and to research, that we make attempts and trials in order to gain facts from which the mind, through reasoning, may draw knowledge or instruction.

Speaking in the abstract, when we say, "relying on observation and gaining experience," we mean that observation is the mind's support in reasoning, and experience the mind's support in deciding, or still better, the fruit of exact reasoning applied to the interpretation of facts.

Observation, then, is what shows facts; experiment is what teaches about facts and gives experience in relation to anything.

An Introduction to the Study of Experimental Medicine (p. 11)

Bohr, Niels
The great extension of our experience in recent years has brought to light the insufficiency of our simple mechanical conceptions and, as a consequence, has shaken the foundation on which the customary interpretation of observation was based . . .

Atomic Theory and Description of Nature
Introductory Survey (p. 2)

Box, G.E.P.
To find out what happens to a system when you interfere with it you have to interfere with it (not just passively observe it).

Technometrics
Use and Abuse of Regression (p. 629)
Volume 8, Number 4, November 1966

Darwin, Charles
Oh, he is a good observer, but he has no power of reasoning!

The Life and Letters of Charles Darwin
Volume I
Mental Qualities (p. 82)

Eddington, Sir Arthur Stanley
I hope I shall not shock the experimental physicists too much if I add that it is also a good rule not to put overmuch confidence in the observational results that are put forward until they have been confirmed by theory . . .

New Pathways in Science (p. 211)

Einstein, Albert
A man should look for what is, and not for what he thinks should be . . .

Quoted by Peter Michelmore in
Einstein, profile of the man (p. 20)

It is the theory which decides what we can observe.

Quoted by Werner Heisenberg in
Physics and Beyond: Encounters and Conversations (p. 77)

Greer, Scott
. . . the link between observation and formulation is one of the most difficult and crucial in the scientific enterprise. It is the process of interpreting our theory or, as some say, of "operationalizing our concepts". Our creations in the world of possibility must be fitted in the world of probability; in Kant's epigram, "Concepts without precepts are empty". It is also the process of relating our observations to theory; to finish the epigram, "Precepts without concepts are blind".

The Logic of Social Inquiry (p. 160)

Hooke, Robert
The truth is, the Science of Nature has been already too long made only a work of the *Brain* and the *Fancy*: It is now high time that it should return to the plainness and soundness of Observations on *material* and *obvious* things.

Micrographia
Preface

Hutten, Ernest H.
. . . certain conditions under which the observable thing is perceived are tacitly assumed . . . for the possibility that we deal with hallucinations or a dream can never be excluded.

The Language of Modern Physics (p. 51)

Jonson, Ben
I do love to note and to observe.

Volpone
Act II, scene 1

Longair, Malcolm
Although by now a large amount of observational material is available, the implications of these observations are far from clear.

Contemporary Physics
Quasi-Stellar Radio Sources (p. 357)
Volume 8, Number 4, 1967

Lyttleton, R.A.
Observations are meaningless without a theory to interpret them.

Quoted by Charles-Albert Reichen in
A History of Astronomy (p. 88)

O'Neil, W.M.
It urges the scientist, in effect, not to take risks incurred in moving far from the facts. However, it may properly be asked whether science can be undertaken without taking the risk of skating on the possibly thin ice of supposition. The important thing to know is when one is on the more solid ground of observation and when one is on the ice.

Fact and Theory
Chapter 8 (p. 154)

Poincaré, Henri
. . . to observe is not enough. We must use our observations, and to do that we must generalize.

The Foundations of Science
Science and Hypothesis
Hypothesis in Physics (p. 127)

Pope, Alexander
To observations which ourselves we make,
We grow more partial for th' observer's sake.

The Complete Poetical Works of POPE
Moral Essays
Epis. I, l. 11–12

Popper, Karl R.
Some scientists find, or so it seems, that they get their best ideas when smoking; others by drinking coffee or whiskey. Thus there is no reason why I should not admit that some may get their ideas by observing or by repeating observations.

Realism and the Aim of Science (p. 36)

Seuss, Dr.
You will see something new.
Two things. And I call them
Thing One and Thing Two.

The Cat in the Hat (p. 33)

Swift, Jonathan
That was excellently observ'd, say I, when I read a Passage in an Author, where his Opinion agrees with mine. When we differ, there I pronounce him to be mistaken.

Satires and Personal Writings
Thoughts on Various Subjects

Wheeler, John A.
May the universe in some strange sense be "brought into being" by the participation of those who participate? . . . the vital act is the act of participation. "Participator" is the incontrovertible new concept given by quantum mechanics. It strikes down the term "observer" of classical theory, the man who stands safely behind the thick glass wall and watches what goes on without taking part. It can't be done, quantum mechanics says.

Gravitation (p. 1273)

Whitehead, Alfred North
'Tis here, 'tis there, 'tis gone.

An Introduction to Mathematics
Chapter 1 (p. 1)

OBSERVATORY

Bierce, Ambrose
OBSERVATORY, *n.* A place where astronomers conjecture away the guesses of their predecessors.

The Enlarged Devil's Dictionary

Cerf, Bennett
Some weeks later the Einsteins were taken to the Mt. Wilson Observatory in California. Mrs. Einstein was particularly impressed by the giant telescope. "What on earth do they use it for?" she asked. Her host explained that one of its chief purposes was to find out the shape of the universe. "Oh," said Mrs. Einstein, "my husband does that on the back of an envelope."

Try and Stop Me
On the telescope (p. 163)

Russell, Henry Norris
The good spectroscopist—to parody the old jest—might perhaps be permitted to go, when he died, instruments and all, and set up an observatory on the moon.

Scientific American
Where Astronomers Go When They Die (p. 112)
Volume 149, Number 3, September 1933

OCCAM'S RAZOR

Chekhov, Anton Pavlovich
It is unfortunate that we try to solve the simplest questions cleverly, and therefore make them unusually complicated. We should seek a simple solution.

Note-Book of Anton Chekhov (p. 20)

Gettings, Fres
Simplex sigillum veri
Cut causes, be merry
Slash 'em and dock 'em
Said William of Ockham
Wiping his razor
On the sleeve of his blazer

Quoted by Renée Haynes in
Times Literary Supplement
Signs of Secrecy (p. 688)
June 19, 1981

Jeans, Sir James Hopwood
When two hypotheses are possible, we provisionally choose that which our minds adjudge to be the simpler, on the supposition that this is the more likely to lead in the direction of the truth. It includes as a special case the principle of Occam's razor—*entia non multiplicanda praeter necessitatem*.

Physics and Philosophy (p. 183)

Entia non sunt multiplicanda praeter necessistatem.
[We must not assume the existence of any entity until we are compelled to do so.]

The Mysterious Universe
Relativity and the Ether (p. 115)

OPTICS

Day, Roger E.
I wish I were a crystal lens,
With aplanatic face,
And lived at Number Seven Ten,
Illumination Place,
City of Glass.

The Physics Teacher
Fantasy of Glass (p. 288)
Volume 3, Number 6, September 1965

Marton, Ladislaus
"Electron optics I believe,'
He often gravely said,
'Concerns a branch of knowledge
That is way above my head, . . . "

American Scientist
Alice in Electronland (p. 251)
Volume 31, Number 3, July 1943

Unknown
Fast into Slow: Dives down Below
Slow into Fast: Takes the Upward Path

Source unknown

ORDER

Davies, Paul Charles William

The universe contains vastly more order than Earth-life could ever demand. All those distant galaxies, irrelevant for our existence, seem as equally well ordered as our own.

Quoted by Eugene F. Mallove in
The Quickening Universe (p. 61)

Huntington, E.V.

The fundamental importance of the subject of order may be inferred from the fact that all the concepts required in geometry can be expressed in terms of the concept of order alone.

The Continuum
Introduction (p. 2)

Picard, Émile

We no longer pretend to be able to grasp reality in a physical theory; we see in it rather an analytic or geometric mold useful and fertile for a tentative representation of phenomena, no longer believing that the agreement of a theory with experience demonstrates that the theory expresses the reality of things. Such statements have sometimes seemed discouraging; we ought rather to marvel that, with representations of things more or less distant and discolored, the human spirit has been able to find its way through the chaos of so many phenomena and to derive from scientific knowledge the ideas of beauty and harmony. It is no paradox to say that science puts order, at least tentative order, into nature.

Quoted by Lucienne Felix in
The Modern Aspect of Mathematics (p. 31)

Russell, Bertrand

Dimensions, in geometry, are a development of order. The conception of a limit, which underlies all higher mathematics, is a serial conception. There are parts of mathematics which do not depend upon the notion of order, but they are very few in comparison with the parts in which this notion is involved.

Introduction to Mathematical Philosophy
The Definition of Order (p. 29)

The notion of continuity depends upon that of order, since continuity is merely a particular type of order.

Mysticism and Logic
Mathematics and Metaphysics (p. 91)
Chapter V

HAPPY BIRTHDAY TO MMMMMmm*f...!*

Consider a pair of twins. Immediately after birth they are separated . . .

Alfred Schild – (See p. 187)

PARADOX

Bohr, Niels
How wonderful that we have met with a paradox. Now we have some hope of making progress.

Quoted by L.I. Ponomarev in
The Quantum Dice (p. 75)

Carvello's Paradox
Imagine that the mysterious Dr. Carvello points a flashlight in your direction and turns it on precisely at midnight. At 11:59 however, a full minute before he actually turns on his light, you put on a special pair of green sunglasses and are able to see the flash of light 60 seconds before it was sent. Explaining how such anticipatory sunglasses are supposed to work is the gist of Carvello's paradox.

Discussed in Nick Herbert's
Faster Than Light (p. 72)

Olbers' Paradox
Consider any large spherical shell centered on the earth. Within this shell, the amount of light produced by stars can be calculated. Then consider a shell of twice the radius. Within this shell, the stars are on the average only one-quarter as bright, but there are 4 times as many of them, and so they would make a similar contribution to the light of the night sky. For each doubling of the radius, the amount of light received on the earth is doubled, and so the night sky must double in brightness. Continuing this argument indefinitely, we find that, as we consider larger and larger shells, the night sky continues to increase in brightness without limit. Yet the night sky is very dark . . .

The Big Bang (p. 55)

Russell, Bertrand
Although this may seem a paradox; all exact science is dominated by the idea of approximation.

Quoted by Jefferson Hane Weaver in
The World of Physics
Volume II (p. 22)

Schild, Alfred
Consider a pair of twins. Immediately after birth they are separated. One of them, the first one, remains on earth, the second on is put in a rocket ship and flown to Alpha Centauri at a pretty high speed, 99% that of light. Alpha Centauri is the nearest star; it is about four light-years away from us. As soon as the second twin gets to Alpha Centauri, he turns around and flies back to earth at the same high speed. When the two twins meet again, the first one, the one who stayed behind on earth, will be *eight years old* . . . he will be able to talk quite well and read a little bit. He may have finished second grade and be about to enter third. The second twin, the one who took the journey, on his return will be approximately *one year old* . . .. He will still need diapers, he will be barely able to walk, and he won't be able to talk much.

American Mathematics Monthly
The Clock Paradox in Relativity Theory (p. 1)
Volume 66, Number 1, January 1959

Schrödinger, Erwin
Attention had recently (A. Einstein, B. Podolsky, and N. Rosen, *Phys. Rev.* **47** (1935) 777) been called to the obvious but very disentangling measurement to one system, the representative obtained for the other system is by no means independent of the particular choice of observations which we select for that purpose and which by the way are entirely arbitrary. It is rather discomforting that the theory should allow a system to be steered or piloted into one or the other type of state at the experimenter's mercy in spite of his having no access to it.

Proceedings of the Cambridge Philosophical Society
Volume 11, 1935 (p. 555)

Schrödinger's Cat
Suppose a cat is confined in a sealed room along with a geiger counter sitting beside an occasional source of radioactivity. If the geiger counter records one of these (for all practical purposes) random decays within an hour, then it triggers the release of poisonous gas into the room which quickly kills the cat. If no atom decays in the allowed time then the cat survives. The experiment ends when we look into the room after an hour to see if the cat is alive or dead. Schrödinger claims that, according to the Copenhagen interpretation of quantum mechanics, before we look inside the room the cat possesses a wave function which describes it as existing

in some mixture of the definite states "dead" and "alive" in which it can be found after the act of looking at the cat determines what state it is in. When and where does the mixed-up, half-dead cat state change from being neither dead nor alive into one or the other? Who collapses the cat's wave function; is it the cat, the geiger counter, or the physicist? Or does quantum theory simply not apply to "large" complicated objects, even if they are composed of smaller ones to which it does apply.

Quoted by John D. Barrow in
The World within the World (p. 152)

Shimony, A.
I hope that the rigor and beauty of the argument of EPR is apparent. If one does not recognize how good an argument it is—proceeding rigorously from premises which are thoroughly reasonable—then one does not experience an adequate intellectual shock when one finds out that the experimental evidence contradicts their conclusions. This shock should be as great as the one experienced by Frege when he read Russell's theoretical paradox and said, "Alas, arithmetic totters!"

Quoted by Franco Selleri in
Quantum Mechanics versus Local Realism (p. 19)

Smith, E.E.
With sufficient knowledge, any possible so-called paradox can be resolved.

Masters of the Vortex (p. 110)

Thomson, William [Lord Kelvin]
In science there are no paradoxes.

Quoted by S.P. Thompson in
The Life of William Thomson, Baron Kelvin of Largs (p. 833)

PARTICLES

Fermi, Enrico
If I could remember the names of all these particles, I'd be a botanist.

Quoted by A. Zee in
Fearful Symmetry (p. 168)

Hein, Piet
Nature, it seems is the popular name
for milliards and milliards and milliards
of particles playing their infinite game
of billiards and billiards and billiards.

Grooks II
Atomyriades

Heisenberg, Werner
The mathematically formulated laws of quantum theory show clearly that our ordinary intuitive concepts cannot be unambiguously applied to the smallest particles. All the words or concepts we use to describe ordinary physical objects, such as position, velocity, color, size, and so on, become indefinite and problematic if we try to use them of elementary particles.

Across the Frontiers (p. 114)

In the light of quantum theory these elementary particles are no longer real in the same sense as objects of daily life, trees or stones, but appear as abstractions derived from the real material of observation in the true sense.

On Modern Physics
Philosophical Problems (p. 13)

Hoffmann, Banesh

"Daddy," she says, "which came first, the chicken or the egg?"

"Yes!"

"Daddy, is it a wave or a particle?"

"Yes."

"Daddy, is the electron here or is it there?"

"Yes."

"Daddy, do scientists really know what they are talking about?"

"Yes!"

The Strange Story of the Quantum (pp. 157–8)

Young, Edward

As Particles, as Atoms ill-perceiv'd . . .

Night Thoughts
Night VI, l. 187

I think physicists are the Peter Pans of the human race. They never grow up, and they keep their curiosity.

I.I. Rabi – (See p. 202)

PENDULUM

Eco, Umberto
That was when I saw the Pendulum.

The sphere, hanging from a long wire set into the ceiling of the choir, swayed back and forth with isochronal majesty.

I knew—but anyone could have sensed it in the magic of that serene breathing—that the period was governed by the square root of the length of the wire and by π, that number which, however irrational to sublunar minds, through a higher rationality binds the circumference and diameter of all possible circles. The time it took the sphere to swing from end to end was determined by an arcane conspiracy between the most timeless of measures: the singularity of the point of suspension, the duality of the plane's dimensions, the triadic beginning of π, the secret quadratic nature of the root, and the unnumbered perfection of the circle itself.

Foucault's Pendulum
Chapter 1 (p. 3)

The ideal pendulum consists of a very thin wire, which will not hinder flexion and torsion, of length L, with the weight attached to its barycenter. For a sphere, the barycenter is the center; for the human body, it is a point 0.65 of the height, measured from the feet. If the hanged man is 1.70 m tall, his barycenter is located 1.10 m from his feet, and the length L includes this distance. In other words, if the distance from the man's head to neck is 0.60 m, the barycenter $1.70 - 1.10 = 0.60$ m from his head, and $0.60 - 0.30 = 0.30$ m from his neck.

The period of the pendulum, discovered by Huygens, is given by:

$$T \text{ (seconds)} = \frac{2\pi}{\sqrt{g}} \sqrt{L} \tag{1}$$

Where L is the length in meters, $\pi = 3.1415972 \ldots$ and $g = 9.8 \text{ m/sec}^2$. Thus (1) gives:

$$T = \frac{2 \times 3.1415972}{\sqrt{9.8}}\sqrt{L} = 2.0079\sqrt{L}$$

or, more or less:

$$T = 2\sqrt{L}.$$

Note: T is independent of the weight of the hanged man. (In God's eyes all men are equal . . .)

Foucault's Pendulum
Chapter 114

Graham, L.A.
Rock-a-bye baby in the tree top,
As a compound pendulum, you are a flop.
Your center of percussion is safe and low,
As one may see when the wind doth blow.
Your frequency of vibration is pretty small,
Frankly, I don't think you'll fall at all.

Ingenious Mathematical Problems and Methods
Mathematical Nursery Rhyme No. 2

PHOTON

Jespersen, James
Fitz-Randolph, Jane
We can think of the photons as being like a shower of snowballs flying back and forth between the two electrons. And like the opponents in a snowball fight, the electrons retreat from each other under the assault of the photons.

Quoted by Robert G. Colodny (Editor) in
From Quarks to Quasars (p. 125)

Roberts, Michael
While I, maybe, precisely seize
The elusive photon's properties
In α's and δ's, set in bronze-
bright vectors, grim quaternions.

The New Statesman
Miscellany (p. 418)
Notes on θ, ϕ, and ψ
Saturday March 23, 1935
Volume 9, Number 213 (New Series)

Rucker, Rudy
A photon is a wavy yet solid little package that can zip through empty space without the benefit of any invisible jelly vibrating underfoot.

The Fourth Dimension
How to get there from here (p. 73)

PHYSICIST

Bergmann, P.
In many aspects, the theoretical physicist is merely a philosopher in a working suit.

Quoted by Jean-Pierre Luminet in
Black Holes (p. 51)

Birkhoff, George David
It is to be hoped that in the future more and more theoretical physicists will command a deep knowledge of mathematical principles; and also that mathematicians will no longer limit themselves so exclusively to the aesthetic development of mathematical abstractions.

American Scientist
Mathematical Nature of Physical Theories (p. 286)
Volume 31, Number 4, October 1943

Boltzmann, Ludwig
$S = k \log w$

Carved on Boltzmann's gravestone

Brillouin, Léon
It is impossible to study the properties of a single mathematical trajectory. The physicist knows only bundles of trajectories, corresponding to slightly different initial conditions.

Quoted by John D. Barrow in
The World within the World (p. 277)

Burroughs, William S.
No atomic physicist has to worry, people will always want to kill other people on a mass scale.

The Adding Machine
A Word to the Wise Guy

Davies, Paul Charles William
Brown, Julian R.
. . . physicists, like theologians, are wont to deny that any system is in principle beyond the scope of their subject.

Superstrings: A Theory of Everything (p. 1)

DiCurcio, Robert
It Is an Ancient Physiker,
And he stoppeth one of three.
"By thy long gray beard and glittering eye,
Now Wherefore stopp'st thou me?"

The Physics Teacher
Physics inspires the Muses
The Rime of the Ancient Physiker (p. 634)
Part I of VII
Volume 16, Number 9, December 1978

Duhem, Pierre
The watchmaker to whom one gives a watch that does not run will take it all apart and will examine each of the pieces until he finds out which one is damaged. The physician to whom one presents a patient cannot dissect him to establish the diagnosis. He has to guess the seat of the illness by examining the effect on the whole body. The physicist resembles a doctor, not a watchmaker.

Revue des Questions Scientifiques
Quelques réflexions au sujet de la physique expérimentale

. . . *if the aim of physical theories is to explain experimental laws, theoretical physics is not an autonomous science; it is subordinate to metaphysics.*

The Aim and Structure of Physical Theory (p. 10)

Dürrenmatt, Friedrich
Dear Mobius. You have visitors. Now leave your physicist's lair for a moment and come in here.

The Physicists
Act One (p. 37)

It's ludicrous. Here we have hordes of highly paid physicists in gigantic state-supported laboratories working for years and years and years vainly trying to make some progress in the realm of physics while you do it quite casually at your desk in this madhouse.

The Physicists
Act Two (p. 75)

Dyson, Freeman J.
Theoretical physicists are accustomed to living in a world which is removed from tangible objects by two levels of abstraction. From tangible atoms we move by one level of abstraction to invisible fields and particles. A second level of abstraction takes us from fields and particles to the symmetry-groups by which fields and particles are related. The superstring theory takes us beyond symmetry-groups to two further levels of abstraction. The third level of abstraction is the interpretation of symmetry-groups in terms of states in ten-dimensional space–time. The fourth level is the world of the superstrings by whose dynamical behavior the states are defined.

Infinite In All Directions (p. 18)

Eddington, Sir Arthur Stanley
Life would be stunted and narrow if we could feel no significance in the world around us beyond that which can be weighed and measured with the tools of the physicist or described by the metrical symbols of the mathematician.

Quoted by Arthur Beiser in
The World of Physics
Introduction

"Wheresoever the carcass is, there will the eagles be gathered together", and where the symbols of the mathematical physicists flock, there presumably is some prey for them to settle on, which the plain man at least will prefer to call by a name suggestive of something more than passive emptiness.

New Pathways in Science (p. 39)

To the pure geometer the radius of curvature is an incidental characteristic—like the grin of the Cheshire cat. To the physicist it is an indispensable characteristic. It would be going too far to say that to the physicist the cat is merely incidental to the grin. Physics is concerned with interrelatedness such as the interrelatedness of cats and grins. In this case the "cat without a grin" and the "grin without a cat" are equally set aside as purely mathematical phantasies.

The Expanding Universe
The Universe and the Atom
Section III (pp. 103–4)

Einstein, Albert
Not everyone is as fortunate as Christ. To sacrifice yourself and do some good, that takes luck.

Quoted by Leo Szilard in
Leo Szilard: His Version of the Facts (p. 12)

If you want to find out anything from the theoretical physicists about the methods they use, I advise you to stick closely to one principle: Don't listen to their words, fix your attention on their deeds.

Idea and Opinion
On the Method of Theoretical Physics

The supreme task of the physicist is to arrive at those universal elementary laws from which the cosmos can be built up by pure deduction.

The World As I See It (p. 22)

The whole of science is nothing more than a refinement of everyday thinking. It is for this reason that the critical thinking of the physicist cannot possibly be restricted to the examination of the concepts of his own specific field. He cannot proceed without considering critically a much more difficult problem, the problem of analyzing the nature of everyday thinking.

Out of My Later Years
Physics and Reality
Section 1

Feynman, Richard P.
For those who want some proof that physicists are human, the proof is in the idiocy of all the different units which they use for measuring energy.

The Character of Physical Law
Chapter 3 (p. 75)

The limited imagination of physicists: When we see a new phenomenon we try to fit it into the framework we already have . . ... It's not because Nature is *really* similar; it's because the physicists have only been able to think of the same damn thing, over and over again.

QED (p. 149)

Foster, G.C.
. . . from the very outset of his investigations the physicist has to rely constantly on the aid of the mathematician, for even in the simplest cases, the direct results of his measuring operations are entirely without meaning until they have been submitted to more or less of mathematical discussion.

Nature
Section A
Mathematical and Physical
Opening Address (p. 312)
Volume 16, Number 407, August 16, 1887

Gamow, George

Now, Physicists, take warning,
  Observe this sober test . . .
When new fleas are a-borning
  Make sure they're fully dressed!

Thirty Years that Shook Physics (p. 193)

. . . that handsome, hearty British lord
We knew as Ernest Rutherford.
New Zealand farmer's son by birth,
He never lost the touch of earth;
His booming voice and jolly roar
Could penetrate the thickest door,
But if to anger he inclined
You should have heard him speak his mind
In living language of the land
That anyone could understand!

One day George Gamow, as his guest,
By Rutherford was so addressed
At tea in honour of Niels Bohr
(Of whom you may have heard before).
The men talked golf, and cricket too;
The ladies gushed, as ladies do,
About a blouse, a sash, a shawl—
And Bohr grew weary of it all.
"Gamow," he said, "I see below
Your motorcycle. Will you show
Me how it works? Come on, let's run!
This party isn't any fun."
So to the motorcycle Bohr,
With Gamow running after, tore.
Gamow explained the this and that
And Bohr, who on the saddle sat,
Took off to skim along the Backs,
A threat to humans, beasts and hacks,
But though he started full and strong
He didn't sit it out for long.
No less than fifty yards ahead
He killed the nervous engine dead
And, turning wildly as he slowed,
Stopped traffic up and down Queen's Road.

While Gamow, rushing to the fore,
Was doing what he could for Bohr
Who should like Jove himself appear

But Rutherford. In Gamow's ear
He thundered: "Gamow! If once more
You give that buggy to Niels Bohr
To snarl up traffic with, or wreck,
I swear I'll break your bloody neck!"

Biography of Physics (pp. 221–2)

Gibbs, J. Willard

A mathematician may say anything he pleases, but a physicist must be at least partially sane.

Quoted by R.B. Lindsay in
The Scientific Monthly
On the Relation of Mathematics and Physics (p. 456)
December 1944

Goethe, Johann Wolfgang von

. . . cement, patch-up, and glue together, as witchdoctors do, the Newtonian doctrine, so that it could, as an embalmed corpse, preside in the style of ancient Egyptians, at the drinking bouts of physicists.

Quoted by Stanley L. Jaki in
American Journal of Physics
Goethe and the Physicists (§211, p. 198)
Volume 37, Number 2, February 1969

Heisenberg, Werner

The physicist may be satisfied when he has the mathematical scheme and knows how to use it for the interpretation of the experiments. But he has to speak about his results also to non-physicists who will not be satisfied unless some explanation is given in plain language. Even for the physicist the description in plain language will be the criterion of the degree of understanding that has been reached.

Physics and Philosophy (p. 168)

Hoffmann, Banesh

They could but make the best of it, and went around with woebegone faces sadly complaining that on Mondays, Wednesdays and Fridays they must look on light as a wave; on Tuesdays, Thursdays and Saturdays, as a particle. On Sundays they simply prayed.

The Strange Story of the Quantum (p. 42)

Kac, Mark

. . . to quote a statement of Poincaré, who said (partly in jest no doubt) that there must be something mysterious about the normal law since mathematicians think it is a law of nature whereas physicists are convinced that it is a mathematical theorem.

Statistical Independence in Probability Analysis and Number Theory
Chapter 3, The Normal Law (p. 52)

Krutch, Joseph Wood
Electronic calculators can solve problems which the man who made them cannot solve; but no government-subsidized commission of engineers and physicists could create a worm . . .

The Twelve Seasons
March (p. 184)

Kuhn, Thomas S.
Looking at a contour map, the student sees lines on paper, the cartographer a picture of a terrain. Looking at a bubble-chamber photograph, the student sees confused and broken lines, the physicist a record of familiar subnuclear events. Only after a number of such transformations of vision does the student become an inhabitant of the scientist's world.

The Structure of Scientific Revolutions
Chapter X (p. 111)

Ladenburg, Rudolf
There are two kinds of physicists in Berlin: on the one hand was Einstein, and on the other all the rest.

Einstein: A Centenary Volume (p. 125)

Lichtenberg, Georg Christoph
The myths of the physicists.

Lichtenberg: Aphorisms & Letters
Aphorisms (p. 57)

March, Robert H.
Physicists are not regular fellows—and neither are poets. Anyone engaged in an activity that makes considerable demands on both the intellect and the emotions is not unlikely to be a little bit odd.

Physics for Poets (p. 1)

Marcus, Adrianne
Let others lie about the universe,
make visible worlds. I am the keeper
of particles, custodian of stray
atoms.

In Steve Rasnic Tem's (Editor)
The Umbral Anthology of Science Fiction Poetry
The Physicist's Purpose

Newman, James R.
In this century the professional philosophers have let the physicists get away with murder. It is a safe bet that no other group of scientists could have passed off and gained acceptance for such an extraordinary principle as complementary, nor succeeded in elevating indeterminacy to a universal law.

Scientific American
Book Review (p. 116)
Volume 198, Number 1, January 1958

Nietzsche, Friedrich
We must be physicists in order . . . to be creative since so far codes of values and ideals have been constructed in ignorance of physics or even in contradiction to physics.

The Gay Science
Aphorism 335

Oppenheimer, J. Robert
In some sort of crude sense which no vulgarity, no humor, no overstatement can quite extinguish, the physicists have known sin, and this is a knowledge which they cannot lose.

Time
Expiation (p. 94)
Volume 51, Number 8, 23 February 1948
See also
The Open Mind (p. 88)

Petroski, Henry
Embedded in a matrix of mistakes
And slips, his next equation lies
About its symmetry. Among the signs
Of exercise and bold heuristic thrusts
Of algebra and calculus, it takes
His magic mirror mind to recognize
A juxtaposition that unifies
His theory of another universe.

Extracting the law from the accidents,
He calls it Theorem and proceeds to prove
It logically follows from stronger laws.
He makes some definitions and extends
The theorem more and marvels at the rules
His universe follows, effect from cause.

Southern Humanities Review
The Mathematical Physicist (p. 184)
Volume 8, Number 2, Spring 1974

Pierce, C.S.

If hypotheses are to be tried haphazard, or simply because they will suit certain phenomena, it will occupy the mathematical physicists of the world say half a century on the average to bring each theory to the test, and since the number of possible theories may go up into the trillion, only one of which can be true, we have little prospect of making further solid additions to the subject in our time.

The Monist
The Architecture of Theories (p. 164)
Volume 11, Number 2, January 1891

Poincaré, Henri

"Nothing but facts are of importance. John Lackland passed by here. Here is something that is admirable. Here is a reality for which I would give all the theories in the world." Carlyle was a fellow countryman of Bacon; but Bacon would not have said that. That is the language of the historian. The physicist would say rather: "John Lackland passed by here; that makes no difference to me, for he never will pass this way again."

The Foundations of Science
Science and Hypothesis
Hypotheses in Physics (p. 128)

Rabi, I.I.

I think physicists are the Peter Pans of the human race. They never grow up, and they keep their curiosity.

In Jeremy Bernstein's
Experiencing Science (p. 102)

Robinson, Howard A.

Because physicists are a small group, they often suffer in many ways from psychoses similar to those found in political minorities. In an effort to keep their own individuality they feel it necessary to resist pressure from the outside and the result is . . . that a group of physicists tend to behave like an amoeba.

Physics Today
The Challenge of Industrial Physics (p. 7)
Volume 1, Number 2, June 1948

Rogers, Eric

The physicist who does not enjoy watching a dime and a quarter drop together has no heart.

Astronomy for the Inquiring Mind (p. 4)

Schrödinger, Erwin

To-day there are not a few physicists who, like Kirchoff and Mach, regard the task of physical theory as being merely a mathematical description (*as economical as possible*) of the empirical connections between observable quantities, i.e., it is a description which reproduces the connection, as far as possible, without the intervention of unobservable elements.

Collected Papers on Wave Mechanics (p. 58)

Standen, Anthony
Physicists, being in no way different from the rest of the population, have short memories for what is inconvenient.

Science is a Sacred Cow (p. 68)

Strutt, John William
The different habits of mind of the two schools of physicists sometimes lead them to the adoption of antagonistic views on doubtful and difficult questions. The tendency of the purely experimental school is to rely almost exclusively upon direct evidence, even when it is obviously imperfect, and to disregard arguments which they stigmatize as theoretical. The tendency of the mathematician is to over-rate the solidity of his theoretical structures, and to forget the narrowness of the experimental foundation upon which many of them rest.

Life of John William Strutt (p. 132)

Thomson, Sir J.J.
There is a school of mathematical physicists which objects to the introduction of ideas which do not relate to things which can actually be observed and measured . . . .. I hold that if the introduction of a quantity promotes clearness of thought, then even if at the moment we have no means of determining it with precision, its introduction is not only legitimate but desirable. The immeasurable of today may be the measurable of tomorrow.

Quoted by John D. Barrow in
The World within the World (p. 97)

Unknown
We seek, we study, and we stare
At particles that weren't quite there.

Quoted by H. Arthur Klein in
The World of Measurements
Song for a High-Energy Physicist (p. 180)

There was once a physicist named Bohr
Who said on Pigalle to a whore:
  "I'm full of $h\nu$
  How about a good screw?
Why, that's what my Fulbright is for!"

The New Limerick
2574

The only distinction between physicists and engineers is the physicists have more questions than answers while engineers have more answers than questions.

Source unknown

THE PHYSICISTS' BILL OF RIGHTS

We hold these postulates to be intuitively obvious, that all physicists are born equal, to a first approximation, and are endowed by their creator with certain discrete privileges, among them a mean rest life, n degrees of freedom, and the following rights which are invariant under all linear transformations:

1. To approximate all problems to ideal cases.

2. To use order of magnitude calculations whenever deemed necessary (i.e., whenever one can get away with it).

3. To use the rigorous method of "squinting" for solving problems more complex than the addition of positive real integers.

4. To dismiss all functions which diverge as "nasty" and "unphysical."

5. To invoke the uncertainty principle when confronted by confused mathematicians, chemists, engineers, psychologists, dramatists, and other lower scientists.

6. When pressed by non-physicists for an explanation of (4) to mumble in a sneering tone of voice something about physically naive mathematicians.

7. To equate two sides of an equation which are dimensionally inconsistent, with a suitable comment to the effect of, "Well, we are interested in the order of magnitude anyway."

8. To the extensive use of "bastard notations" where conventional mathematics will not work.

9. To invent fictitious forces to delude the general public.

10. To justify shaky reasoning on the basis that it gives the right answer.

11. To cleverly choose convenient initial conditions, using the principle of general triviality.

12. To use plausible arguments in place of proofs, and thenceforth refer to these arguments as proofs.

13. To take on faith any principle which seems right but cannot be proved.

Quoted on the Internet

A mathematician and a physicist agree to a psychological experiment. The mathematician is put in a chair in a large empty room and a beautiful naked woman is placed on a bed at the other end of the room. The psychologist explains, "You are to remain in your chair. Every five minutes, I will move your chair to a position halfway between its current location and the woman on the bed." The mathematician looks at the psychologist in disgust. "What? I'm not going to go through this. You know I'll never reach the bed!" And he gets up and storms out. The psychologist makes a note on his clipboard and ushers the physicist in. He explains the situation, and the physicist's eyes light up and he starts drooling. The psychologist is a bit confused. "Don't you realize that you'll never reach her?" The physicist smiles and replies, "Of course! But I'll get close enough for all practical purposes!"

Source unknown

A Princeton plasma physicist is at the beach when he discovers a ancient looking oil lantern sticking out of the sand. He rubs the sand off with a towel and a genie pops out. The genie offers to grant him one wish. The physicist retrieves a map of the world from his car and circles the Middle East and tells the genie, "I wish you to bring peace in this region".

After 10 long minutes of deliberation, the genie replies, "Gee, there are lots of problems there with Lebanon, Iraq, Israel, and all those other places. This is awfully embarrassing. I've never had to do this before, but I'm just going to have to ask you for another wish. This one is just too much for me".

Taken aback, the physicist thinks a bit and asks, "I wish that the Princeton tokamak would achieve scientific fusion energy break-even."

After another deliberation the genie asks, "Could I see that map again?"

Source unknown

A physicist is spending a vacation in Geneva near the headquarters of CERN. He's helping his wife shop, and she decides to have her hair done. She tells him, "Be back at exactly 4:00 to pick me up, no Physics!" As he walks down the avenue, he sees a cute blonde peering under the hood of her car.

"Can I help you?" he asks.

"I'm stalled."

He takes out a Swiss Army knife and fiddles with the engine. "Try it." No luck. He fiddles again. "Try it again." No luck. He fiddles some more. "Try it again."

Vrooooom, vrooooom. Success at last.

She says, "Thank you very much. Oh dear, your hands are covered with grease. I live nearby. You can stop and wash your hands."

He washes his hands. She offers him tea. One thing leads to another, and they jump into bed together. An hour later he jumps out of bed and says, "Look at the time!" He dresses quickly and rushes out the door. Then he stops, goes back in and asks, "Got any chalk?"

"Yeah, in the drawer over there."

He marks the back of his jacket with chalk and rushes to the hairdresser where his wife is waiting with packages and black smoke coming out of her ears. "Where were you?"

"Well, I was walking down the street and saw this attractive blonde whose car wouldn't start. I helped her, got all greasy, and went upstairs to wash my hands. One thing led to another and we jumped into bed . . . "

She says, "Wait a minute; turn around. You liar—you went to CERN and talked physics!"

Source unknown

Wheeler, John A.
The physicist does not have the habit of giving up something unless he gets something better in return.

Quoted by Cecil M. DeWitt and John A. Wheeler in
Battelle Rencontres (p. 261)
1967 Lectures in Mathematics and Physics

White, Stephen
[Physicists] are, as a general rule, highbrows. They think and talk in long, Latin words, and when they write anything down they usually include at least one partial differential and three Greek letters.

Physics Today
A Newsman Looks at Physicists (p. 15)
Volume 1, Number 1, May 1948

Wigner, Eugene
Part of the art and skill of the engineer and of the experimental physicist is to create conditions in which certain events are sure to occur.

Symmetries and Reflections
The Role of Invariance Principles in Natural Philosophy (p. 29)

Zee, Anthony
A gourmet tastes a hollandaise sauce and mutters disapprovingly, "A touch too much lemon, I say." Physicists proceed along the same line, "tasting" the universe to find out what the Ultimate Cook put in.

An Old Man's Toys (p. 99)

PHYSICS

Alvarez, Luis
There is no democracy in physics. We can't say that some second rate guy has as much right to opinion as Fermi.

Quoted by D.S. Greenberg in
The Politics of Pure Science (p. 43)

Bacon, Francis
Physic . . . is situate in a middle term or distance between natural history and metaphysic. For natural history describeth the variety of things; physic the causes, but variable or respective causes; and metaphysic the fixed and constant causes.

Advancement of Learning
Second Book, VII, Section 4

We have no sound notions either in logic or physics; substance, quality, action, passion, and existence are not clear notions; much less weight, levity, density, tenuity, moisture, dryness, generation, corruption, attraction, repulsion, element, matter, form, and the like. They are all fantastical and ill-defined.

Novum Organum
First Book, Section 15

Ball, Walter William Rouse
The advance in our knowledge of physics is largely due to the application to it of mathematics, and every year it becomes more difficult for an experimenter to make any mark in the subject unless he is also a mathematician.

A Short Account of the History of Mathematics (p. 503)

Barrett, John A.
I've studied all the sciences in order alphabetical,
My judgment is, which some of you may find to be heretical,

The field that's really quite abstruse,
The field where all the screws come loose,
The field that's famous for its spoofs, is physics theoretical.

I've taken undergraduate work whose content is forgettable;
And graduate work is gen'rally regarded as regrettable.
The lecturers are all absurd.
A cogent word is never heard.
Insanity afflicts a third in physics theoretical.

We never do experiments; we shun the purely practical.
Our best work's done in getting grants—our budgets are fantastical.
In one respect our motive's pure:
Though funding fails, we still endure—
We make damn sure our job's secure in physics theoretical.

Our scientific breakthroughs are, to say the least, debatable.
We laugh at critics haughtily; our egos are inflatable.
The rest of science goes along,
Because our last defense is strong:
It's hard to prove we're ever wrong in physics theoretical.

Source unknown

Bell, E.T.
Daniel Bernoulli has been called the father of mathematical physics.

Quoted by James R. Newman in
The World of Mathematics
Volume Two (p. 774)

Bergson, Henri
. . . physics is but logic spoiled.

Creative Evolution (p. 320)

Bernoulli, Daniel
. . . it would be better for the true physics if there were no mathematicians on earth.

Quoted in
The Mathematical Intelligencer
Letter to the Editor
Filler (p. 6)
Volume 13, Number 1, Winter 1991

Blackett, P.M.S.
Thus was born the vast modern subject of nuclear physics, which now gives such fertile research problems to so many of the world's physicists and, incidentally, such headaches to so many of the world's statesmen.

In J.B. Birks'
Rutherford at Manchester (p. 104)
Memories of Rutherford

Bohr, Niels
It is wrong to think that the task of physics is to find out how Nature is. Physics concerns what we can say about Nature.

Quoted by Heinz R. Pagels in
The Cosmic Code (p. 85)

Born, Max
It is natural that a man should consider the work of his hands or his brain to be useful and important. Therefore nobody will object to an ardent experimentalist boasting of his measurements and rather looking down on the "paper and ink" physics of his theoretical friend, who on his part is proud of his lofty ideas and despises the dirty fingers of the other.

Experiment and Theory in Physics (p. 1)

The problem of physics is how the actual phenomena, as observed with the help of our sense organs aided by instruments, can be reduced to simple notions which are suited for precise measurement and used for the formulation of quantitative laws.

Experiment and Theory in Physics (pp. 8–9)

Boyle, Robert
I confess, that after I began . . . to discern how useful mathematicks may be made to physicks, I have often wished that I had employed the speculative part of geometry, and the cultivation of the specious Algebra I had been taught very young, a good part of that time and industry, that I had spent about surveying and fortification (of which I remember I once wrote an entire treatise) and other parts of practick mathematicks.

The Usefulness of Mathematicks to Natural Philosophy
Works
Volume 3 (p. 426)

Bragg, Sir William
On Mondays, Wednesdays, and Fridays we teach the wave theory and on Tuesdays, Thursdays, and Saturdays the corpuscular theory.

Scientific Monthly
Electrons and Ether Waves
23rd Robert Boyle Lecture (p. 11)
Oxford, 1921

Bronowski, Jacob
Physics becomes in those years the greatest collective work of art of the twentieth century.

The Ascent of Man
Chapter 10 (p. 328)

Camus, Albert
At the final stage you teach me that this wondrous and multicolored universe can be reduced to the atom and that the atom itself can be reduced to the electron. All this is good and I wait for you to continue. But you tell me of an invisible planetary system in which electrons gravitate around a nucleus. You explain this world to me with an image.

The Myth of Sisyphus
An Absurd Reasoning (p. 15)

Carnap, Rudolf
. . . the facts and objects of the various branches of Science are fundamentally the same kind. For all branches are part of the unified Science, of Physics.

The Unity of Science (p. 101)

Physics originally began as a descriptive macrophysics, containing an enormous number of empirical laws with no apparent connections. In the beginning of a science, scientists may be very proud to have discovered hundreds of laws. But, as the laws proliferate, they become unhappy with this state of affairs; they begin to search for underlying principles.

Quoted by John D. Barrow in
The World within the World (p. 160)

CERN Courier
The main goal of physics is to describe a maximum of phenomena with a minimum of variables.

Quoted by John N. Shive and Robert L. Weber in
Similarities in Physics (p. 213)

Comte, Auguste
The domain of physics is no proper field for mathematical pastimes. The best security would be in giving a geometrical training to physicists, who need not then have recourse to mathematicians, whose tendency is to despise experimental science.

The Positive Philosophy
Book III, Chapter I, Volume I (p. 182)

. . . the education of physicists must be more complicated than that of astronomers.

The Positive Philosophy
Book III, Chapter I, Volume I (p. 183)

Darrow, Karl K.

. . . it does not take an idea so long to become "classical" in physics as it does in the arts.

Bell System Technical Journal
Some Contemporary Advances in Physics V
Electrical Solids (p. 621)
Volume 3, 1924

Davies, Paul Charles William
Brown, Julian R.

No science is more pretentious than physics, for the physicist lays claim to the whole universe as his subject matter.

Superstrings: A Theory of Everything (p. 1)

de Morgan, Augustus

Among the mere talkers so far as mathematics are concerned, are to be ranked three out of four of those who apply mathematics to physics, who, wanting a tool only, are very impatient of everything which is not of direct aid to the actual methods which are in their hands.

Quoted by Robert Graves in
Life of Sir William Rowan Hamilton
Volume 3 (p. 348)

Descartes, René

I accept no principles of physics which are not also accepted in mathematics . . .

Principles of Philosophy
Part III, Principle IV

Dilorenzo, Kirk

Physics is the interrelationship of everything.

The Physics Teacher
What's Physics? (p. 315)
Volume 14, Number 5, May 1976

Duhem, Pierre

A "Crucial Experiment" is impossible in Physics.

The Aim and Structure of Physical Theory (p. 188)

The development of physics incites a continual struggle between "nature that does not tire of providing" and reason that does not wish "to tire of conceiving."

The Aim and Structure of Physical Theory (p. 23)

Dyer, Charles
Physics is the most basic of sciences.
What other science can describe the trajectory
	of a baseball in a vacuum, or need to?
In Experiment One we discovered the latent properties
	of the meter stick.
Verification of gravity was next with the discovery of
	friction in Part B, figure 12-12.
In a year or two we will discover over again
	the ratio of charge to mass of an electron!
(In case anyone forgot.)

The Physics Teacher
Physics (p. 321)
Volume 6, Number 6, September 1968

Dyson, Freeman J.
Physics is littered with the corpses of dead unified field theories.

Quoted by John D. Barrow in
The World within the World (p. 184)

Eddington, Sir Arthur Stanley
In the world of physics we watch a shadowgraph performance of familiar
life. The shadow of my elbow rests on the shadow table as the shadow ink
flows over the shadow paper . . .. The frank realism that physical science
is concerned with a world of shadows is one of the most significant of
recent advances.

The Nature of the Physical World
Introduction (p. xi)

I have not suggested that religion and free will can be deduced from
modern physics . . .

New Pathways in Science (p. 306)

It is impossible to trap modern physics into predicting anything with
perfect determinism because it deals with probabilities from the outset.

Quoted by James R. Newman in
The World of Mathematics
Volume Two (p. 1056)

Sir William Bragg was not overstating the case when he said that we
use the classical theory on Mondays, Wednesdays and Fridays, and the
quantum theory on Tuesdays, Thursdays and Saturdays.

The Nature of the Physical World
Chapter IX
Relation of Classical Laws to Quantum Laws (p. 194)

Distance and duration are the most fundamental terms in physics; velocity, acceleration, force, energy, and so on, all depend on them; and we can scarcely make any statement in physics without direct or indirect reference to them.

> Quoted by Ronald W. Clark in
> *Einstein: The Life and Times* (p. 93)

Edgeworth, Francis Ysidro
In short, Statistics reigns and revels in the very heart of Physics.

> *Journal of the Royal Statistical Society*
> On the Use of the Theory of Probabilities in Statistics Relating to Society (p. 167)
> January 1913

Einstein, Albert
In speaking here of "comprehensibility," the expression is used in its most modest sense. It implies: the production of some sort of order among sense impressions, this order being produced by the creation of general concepts, relations between these concepts and sense experience. It is in this sense that the world of our sense experiences is comprehensible. The fact that it is comprehensible is a miracle.

> *Out of My Later Years*
> Physics and Reality
> Section 1

In the matter of physics, the first lesson should contain nothing but what is experimental and interesting to see. A pretty experiment is in itself often more valuable than twenty formulae extracted from our minds; it is particularly important that a young mind that has yet to find its way about in the world of phenomena should be spared from formulae altogether. In his physics they play exactly the same weird and fearful part as the figures of dates in Universal History.

> *Einstein: A Centenary Volume* (p. 220)

Experience, of course, remains the sole criterion for the serviceability of mathematical constructions for physics, but the truly creative principle resides in mathematics.

> Quoted by Frank Phillip in
> *Modern Science and its Philosophy* (p. 297)

But in physics I soon learned to scent out the paths that led to the depths, and to disregard everything else, all the many things that clutter up the mind, and divert it from the essential. The hitch in this was, of course, the fact that one had to cram all this stuff into one's mind for the examination, whether one liked it or not.

> Quoted by Robert H. March in
> *Physics for Poets* (p. 104)

Today we know that no approach which is founded on classical mechanics and electrodynamics can yield a useful radiation formula.

> Quoted by B.L. van der Waerden in
> *Sources of Quantum Mechanics* (p. 63)
> On the Quantum Theory of Radiation

Reality is the real business of physics.

> Quoted by Nick Herbert in
> *Quantum Reality* (p. 4)

What would physics look like without gravitation?

> Quoted by Jean-Pierre Luminet in
> *Black Holes* (p. 114)

Physics constitutes a logical system of thought which is in a state of evolution, whose basis cannot be distilled, as it were, from experience by an inductive method, but can only be arrived at by free invention.

> *Out of My Later Years*
> Physics and Reality
> Summary

Physics too deals with mathematical concepts; however, these concepts attain physical content only by the clear determination of their relation to the objects of experience.

> *Out of My Later Years*
> The Theory of Relativity (p. 41)

Physics is the attempt at the conceptual construction of a model of the real world and its lawful structure.

> Quoted by G. Holton in
> *Thematic Origins of Scientific Thought* (p. 243)
> Letter of November 28, 1930 to M. Schlick

I have become an evil renegade who does not wish physics to be used on probabilities.

> Quoted by Sachi Sri Kantha in
> *An Einstein Dictionary* (p. 96)

I still lose my temper dutifully about physics. But I no longer flap my wings—I only ruffle my feathers. The majority of fools remain invincible.

> Quoted by Sachi Sri Kantha in
> *An Einstein Dictionary* (p. 96)

Emerson, Ralph Waldo
The axioms of physics translate the laws of ethics.

> *The Works of Ralph Waldo Emerson*
> Volume I
> Nature
> Language (p. 33)

How calmly and genially the mind apprehends one after another the laws of Physics!

> *The Works of Ralph Waldo Emerson*
> Volume I
> Nature
> Discipline (p. 39)

Feynman, Richard P.
In its efforts to learn as much as possible about nature, modern physics has found that certain things can never be "known" with certainty. Much of our knowledge must always remain uncertain. The most we can know is in terms of probabilities.

> *The Feynman Lectures on Physics*
> Volume 1
> Chapter 6-5 (p. 6-11)

Physics is to mathematics what sex is to masturbation.

> Quoted by Lawrence M. Krauss in
> *Fear of Physics* (p. 27)

The fact that I beat a drum has nothing to do with the fact that I do theoretical physics. Theoretical physics is a human endeavor, one of the higher developments of human beings—and this perpetual desire to prove that people who do it are human by showing that they do other things that a few other humans do (like playing bongo drums) is insulting to me. I'm human enough to tell you to go to hell.

> Quoted by James Gleick in
> *Genius: The Life and Science of Richard Feynman*
> Caltec (p. 364)

I often use the analogy of a chess game: one can learn all the rules of chess, but one doesn't know how to play well . . . .. The present situation in physics is as if we know chess, but we don't know one or two rules. But in this part of the board where things are in operation, those one or two rules are not operating much and we can get along pretty well without understanding those rules. That's the way it is, I would say, regarding the phenomena of life, consciousness and so forth.

> Quoted by P.C.W. Davies and J. Brown in
> *Superstrings: A Theory of Everything* (p. 203)

Franklin, W.S.
Physics is the science of the ways of taking hold of things and pushing them.

> Quoted by R.B. Lindsay in
> *The Scientific Monthly*
> The Broad Point of View in Physics (p. 115)
> February 1932

Gamow, George
Can no one laugh? Will no one drink?
I'll teach you physics in a wink . . .

> *Thirty Years that Shook Physics* (p. 190)

Geordi
Suddenly it's like the laws of physics went right out the window.

> *Star Trek: The Next Generation*
> In "True Q"

Hanson, Norwood Russell
Physics is not applied mathematics. It is a natural science in which mathematics can be applied.

> *Patterns of Discovery* (p. 72)

Heinlein, Robert A.
Physics doesn't have to have any use. It just is.

> *Time for the Stars* (p. 138)

Heisenberg, Werner
This assumption is not permissible in atomic physics; the interaction between observer and object causes uncontrollable and large changes in the system being observed, because of the discontinuous changes characteristic of atomic processes.

> *The Physical Principles of the Quantum Theory* (p. 3)

Can nature possibly be as absurd as it seemed to us in these atomic experiments?

> *Physics and Philosophy* (p. 42)

Exact science of the last thirty years derives its special significance from the fact that its different branches, i.e., Astronomy, Physics and Chemistry have been followed back to their common root—atomic physics.

> *Philosophical Problems of Nuclear Science* (p. 27)

Heyl, Paul
Physics is a state of mind.

Quoted by R.B. Lindsay in
The Scientific Monthly
The Broad Point of View in Physics (p. 115)
February 1932

Hilbert, David
Physics . . . is much too hard for physicists.

Quoted by Constance Reid in
Hilbert (p. 127)

Hoyle, Fred
Hoyle, Geoffrey
"In physics," he said, "we plan. We plan months ahead, years ahead . . . You astronomers don't plan, you rush around like a chicken without a head. Observe and observe and observe and all shall be revealed unto you."

The Inferno (p. 77)

Huebner, Jay S.
Physics is what a group of people who call themselves physicists do.

The Physics Teacher
What's Physics? (p. 315)
Volume 14, Number 5, May 1976

Jeans, Sir James Hopwood
Kronecker is quoted as saying that in arithmetic God made the integers and man made the rest; in the same spirit we may perhaps say that in physics God made the mathematics and man made the rest.

Physics and Philosophy
Chapter 1 (p. 16)

. . . physics tries to discover the pattern of events which controls the phenomena we observe. But we can never know what this pattern means or how it originates; and even if some superior intelligence were to tell us, we should find the explanation unintelligible.

Physics and Philosophy
Chapter I (p. 16)

. . . the tendency of modern physics is to resolve the whole material universe into waves, and nothing but waves. These waves are of two kinds: bottled-up waves, which we call matter, and unbottled waves, which we call radiation or light.

The Mysterious Universe
Matter and Radiation (p. 77)

Larrabee, Eric
Some people think that physics was invented by Sir Francis Bacon, who was hit by an apple when he was sitting under a tree one day writing Shakespeare.

Humor from Harper's (p. 89)

Lewis, C.S.
Without a parable modern physics speaks not to the multitudes.

Quoted by John D. Barrow in
The World within the World (p. 238)

Liebson, Morris
"What will I learn here?" you might query.
You'll learn some math and Einstein's Theory.
Ask Teacher for an illustration.
He'll explain, *"It's time dilation.*
Length gets less. Mass gets more.
Time decreases. That's the law.
When things go so very, very fast.
Classical physics is of the past,
And to find what's really true,
We must seek the physics new."
Learning this is lots of fun
In our course called Physics 1.

The Physics Teacher
Physics inspires the Muses (p. 636)
Physics 1
Volume 16, Number 9, December 1978

Mach, Ernst
Physics is experience, arranged in economical order.

Quoted by John N. Shive and Robert L. Weber in
Similarities in Physics
Preface

Maimonides, Moses
. . . he who wishes to attain to human perfection, must therefore first study Logic, next the various branches of Mathematics in their proper order, then Physics, and lastly Metaphysics.

The Guide for the Perplexed
Part I, Chapter XXXIV (p. 46)

Maxwell, James Clerk
. . . a being whose facilities are so sharpened that he can follow every molecule in his course, and would be able to do what is at present impossible to us . . .

Reference to what is called Maxwell's demon
The Theory of Heat

Mayer, Maria Goeppert
Mathematics began to seem too much like puzzle solving. Physics is puzzle solving, too, but of puzzles created by nature, not by the mind of man.

Quoted by J. Dash in
A Life of One's Own (p. 252)

Mencken, H.L.
It is now quite lawful for a Catholic woman to avoid pregnancy by a resort to mathematics, though she is still forbidden to resort to physics and chemistry.

Minority Report: H.L. Mencken's Notebooks
Sample 62

Mohapatra, Rabindra
If you want to do serious physics, sometime you just have to learn it.

As reported by Ernest Barreto
Student, Quantum Field Theory Class 1994

Nietzsche, Friedrich
We must be physicists in order to be creators in that sense,—whereas until now all appreciations and ideals have been based on *ignorance* of physics, or in the contradiction thereto. And therefore, three cheers for physics; and still louder cheers for that which *impels* us thereto—our honesty.

Nietzsche - Schlechta
Volume III
Joyful Wisdom (p. 197)

Noll, Ellis D.
Physics is the science whose treehouse rests on the trunk of immutable physical law.

The Physics Teacher
What's Physics? (p. 315)
Volume 14, Number 5, May 1976

Oman, John
Beauty . . . is the goal of physics as it seeks to construe the order of the universe . . .

The Natural and the Supernatural
Value and Validity (p. 211)

Oppenheimer, J. Robert

The only thing that we can say about the properties of the ultimate particles is that we know nothing whatever about them.

Quoted by Cecilia Helena Payne Gaposchkin in
Introduction to Astronomy (p. 339)

Time and experience have clarified, refined and enriched our understanding of these notions. Physics has changed since then. It will change even more. But what we have learned so far, we have learned well. If it is radical and unfamiliar and a lesson that we are not likely to forget, we think that the future will be only more radical and not less, only more strange and not more familiar, and that it will have its own new insights for the inquiring human spirit.

Quoted by Lucienne Felix in
The Modern Aspect of Mathematics (p. 31)

As you undoubtedly know, theoretical physics—what with the haunting ghosts of neutrinos, the Copenhagen conviction, against all evidence, that cosmic rays are protons, Born's absolutely unquantizable field theory, the divergence difficulties with the positron, and the utter impossibility of making a rigorous calculation of anything at all—is in a hell of a way.

Quoted by Alice Smith and Charles Weiner in
Robert Oppenheimer, Letters and Reflections
Letter to F. Oppenheimer
4 June 1934 (p. 181)

Pais, Abraham

"It was a *wonderful* mess at that time. *Wonderful!* Just great! It was so *confusing*—physics at its best, when everything is confused and you know something important lies just around the corner.

In Robert Crease's
The Second Creation (p. 177)

... the state of particle physics ... is ... not unlike the one in a symphony hall before the start of a concert. On the podium one will see some but not all of the musicians. They are tuning up. Short brilliant passages are heard on some of the instruments; improvisations elsewhere; some wrong notes too. There is a sense of anticipation for the moment when the concert starts.

Physics Today
Particles (p. 28)
Volume 21, Number 2, May 1968

Planck, Max

In endeavoring to claim your attention for a short time, I would remark that our science, Physics, cannot attain its object by direct means, but only gradually along numerous and devious paths, and that therefore a wide scope is provided for the individuality of the worker. One works at one branch, another at another, so that the physical universe with which we are all concerned appears in different lights to different workers.

A Survey of Physics (p. 1)

Modern Physics impresses us particularly with the truth of the old doctrine which teaches that there are realities existing apart from our sense-perceptions, and that there are problems and conflicts where these realities are of greater value for us than the richest treasures of the world of experience.

The Universe in the Light of Modern Physics (p. 107)

Rabi, I.I.
I think physics is infinite. You don't have to try to exhaust it in your generation, or in your lifetime.

Quoted by Jeremy Bernstein in
Experiencing Science (p. 56)

Roberts, Michael
Thomas, E.R.
The most brilliant discoveries in theoretical physics are not discoveries of new laws, but of terms in which the law can be discovered.

Newton and the Origin of Colours (p. 6)

Russell, Bertrand
Physics must be interpreted in a way which tends toward idealism, and perception in a way which tends toward materialism.

The Analysis of Matter
Chapter I (p. 7)

It is obvious that a man who can see, knows things that a blind man cannot know; but a blind man can know the whole of physics.

The Analysis of Matter
Chapter XXXVII (p. 389)

The aim of physics, consciously or unconsciously, has always been to discover what we may call the causal skeleton of the world.

The Analysis of Matter
Chapter XXXVII (p. 391)

Naive realism leads to physics, and physics, if true, shows that naive realism is false. Therefore naive realism, if true, is false; there it is false.

Quoted by John D. Barrow in
The World within the World (p. 144)

Physics is mathematical not because we know so much about the physical world, but because we know so little: it is only its mathematical properties that we can discover.

Quoted by John D. Barrow in
The World within the World (p. 278)

Broadly speaking, traditional physics has collapsed into two portions, truisms and geography.

The ABC of Relativity (p. 222)

Rutherford, Ernest
All physics is either physics or stamp collecting.

Quoted by J.B. Birks
Rutherford at Manchester
Memories of Rutherford (p. 108)

Schaefer, A.
Gershenson, A.
Allersma, M.
A is for ATOM which is really small
B is for BUBBLE CHAMBER where you can see them all
C is for CHARGE which can be quite shocking
D is for DIPOLE gives radiation by oscillating
E is for EINSTEIN who said $E = mc^2$
F is for FEYNMAN whose diagrams make me scared
G is for GRAVITY pulling things down
H is for HYPERFINE STRUCTURE in hydrogen found
I is for INERTIA explaining lethargy
J is for JOULE a unit of energy
K is for KIRCHOFF'S LAWS to get the current right
L is for LASER a really bright light
M is for MAXWELL and his cool equations
N is for NEWTON and his integrations
O is for OPTICS and all that light biz . . .
P is for PHYSICIST which is what daddy (mommy) iz
Q is for QUANTUM and fun with Prof. Yao
R is for RELATIVITY what time is it now?
S is for SEMICONDUCTORS which are really cool
T is for TRANSISTOR a useful tool
U is for UNIFIED FIELD THEORY which remains to be found
V is for VOLTAGE (don't forget to ground!)
W is for WAVE with its myriad effects
X is for X-RAY what else did you expect?
Y is for YOUNG'S DOUBLE SLIT EXPERIMENT (what a mouthful)
Z is for ZEEMAN EFFECT and that is all!

The ABC of Physics

Schrödinger, Erwin
Research in physics has shown beyond the shadow of a doubt that in the overwhelming majority of phenomena whose regularity and invariability have led to the formulation of the postulate of causality, the common element underlying the consistency observed is chance.

Quoted by John D. Barrow in
The World within the World (p. 116)

Scotty
But I canna change the laws of physics, Captain!

To Captain Kirk on *Star Trek*

Shakespeare, William
The labour we delight in physics pain.

Macbeth
Act II, scene III, l. 56

Shirer, W.L.
Modern physics is an instrument of Jewry for the destruction of Nordic Science . . ... True physics is the Creation of the German spirit.

The Rise and Fall of the Third Reich
Chapter 8

Smith, H.J.S.
So intimate is the union between mathematics and physics that probably by far the larger part of the accessions to our mathematical knowledge have been obtained by the efforts of mathematicians to solve the problems set to them by experiment, and to create "for each successive class of phenomena, a new calculus or a new geometry, as the case might be, which might prove not wholly inadequate to the subtlety of nature." Sometimes, indeed, the mathematician has been before the physicists, and it has happened that when some great and new question has occurred to the experimentalist or the observer, he has found in the armory of the mathematician the weapons which he has needed ready made to his hand. But, much oftener, the questions proposed by the physicist have transcended the utmost powers of the mathematics of the time, and a fresh mathematical creation has been needed to supply the logical instrument requisite to interpret the new enigma.

Nature
Presidential Address
British Association for the Advancement of Science (p. 450)
Section A
Volume 8, Number 204, September 25, 1873

Snow, C.P.

I now believe that if I had asked an even simpler question—such as, What do you mean by mass, or acceleration, which is the scientific equivalent of saying, *Can you read?*—not more than one in ten of the highly educated would have felt that I was speaking the same language. So the great edifice of modern physics goes up, and the majority of the cleverest people in the western world have about as much insight into it as their neolithic ancestors would have had.

The Two Cultures (p. 16)

Standen, Anthony

This is the explanation of the extraordinary degree of dullness that pervades the laboratory periods of physics courses, a dullness so acute that for many people it is the bitterest experience of their education.

Science is a Sacred Cow (p. 83)

Stewart, Dugald

According to the doctrine now stated, the highest, or rather the only proper object of physics, is to ascertain those established conjunctions of successive events, which constitute the order of the universe; to record the phenomena which it exhibits to our observations, or which it discloses to our experiments; and to refer these phenomena to their general laws.

Elements of the Philosophy of the Human Mind
Volume 2, Chapter 4, section 1 (p. 524)

Sullivan, J.W.N.

The present tendency of physics is toward describing the universe in terms of mathematical relations between unimaginable entities.

The Bases of Modern Science (p. 226)

Sylvester, J.J.

The object of pure Physic is the unfolding of the laws of the intelligible world; the object of pure Mathematic that of unfolding the laws of human intelligence.

On a Theorem Connected with Newton's Rule
Collected Mathematical Papers
Volume 3 (p. 424)

Unknown
A spaceman and girl in free fall
Obeyed the progenitive call.
  But Newton's Third Rule
  Grabbed hold of his tool,
And shot him across to the wall.

The New Limerick
2597

A young man of Novorossisk
Had a mating procedure so brisk,
  With such superspeed action
  The Lorentz contraction
Foreshortened his prick to a disk.

The New Limerick
2636

If you think, you experience time.
If you feel, you experience energy.
If you intuit, you experience wavelength
If you sense, you experience space.

Quoted by Fred Alan Wolf in
Star Wave (p. 16)

I have a physics teacher,
  I shall not pass,
He feedeth me more than I can learn . . .

The Physics Teacher
A Poem (p. 109)
Volume 14, Number 2, February 1976

If it's green and wiggles, it's biology.
  If it stinks, it's chemistry.
    If it doesn't work, it's physics.

Source unknown

The Euclidean foundation of geometry is to the Gaussian foundation of
geometry as the Newton particle concept of physics is to the Faraday–
Maxwell concept of physics.

Quoted by Howard W. Eves in
Mathematical Circles Squared (p. 36)

IS THERE A SANTA CLAUS?

No known species of reindeer can fly. BUT there are 300,000 species of living organisms yet to be classified, and while most of these are insects and germs, this does not COMPLETELY rule out flying reindeer which only Santa has ever seen.

There are 2 billion children (persons under 18) in the world. BUT since Santa doesn't (appear) to handle the Muslim, Hindu, Jewish and Buddhist children, that reduces the workload to 15% of the total— 378 million according to Population Reference Bureau. At an average (census) rate of 3.5 children per household, that's 91.8 million homes. One presumes there's at least one good child in each.

Santa has 31 hours of Christmas to work with, thanks to the different time zones and the rotation of the earth, assuming he travels east to west (which seems logical). This works out to 822.6 visits per second. This is to say that for each Christian household with good children, Santa has 1/1000th of a second to park, hop out of the sleigh, jump down the chimney, fill the stockings, distribute the remaining presents under the tree, eat whatever snacks have been left, get back up the chimney, get back into the sleigh and move on to the next house. Assuming that each of these 91.8 million stops are evenly distributed around the earth (which, of course, we know to be false but for the purposes of our calculations we will accept), we are now talking about 0.78 miles per household, a total trip of $75\frac{1}{2}$ million miles, not counting stops to do what most of us must do at least once every 31 hours, plus feeding, etc. This means that Santa's sleigh is moving at 650 miles per second, 3,000 times the speed of sound. For purposes of comparison, the fastest man-made vehicle on earth, the Ulysses space probe, moves at a pokey 27.4 miles per second— a conventional reindeer can run, tops, 15 miles per hour.

The payload on the sleigh adds another interesting element. Assuming that each child gets nothing more than a medium-sized Lego set (2 pounds), the sleigh is carrying 321,300 tons, not counting Santa, who is invariably described as overweight. On land, conventional reindeer can pull no more than 300 pounds. Even granting that "flying reindeer" could pull TEN TIMES the normal amount, we cannot do the job with eight, or even nine. We need 214,200 reindeer. This increases the payload— not even counting the weight of the sleigh—to 353,430 tons. Again, for comparison—this is four times the weight of the Queen Elizabeth.

353,000 tons traveling at 650 miles per second creates enormous air resistance - this will heat the reindeer up in the same fashion as space crafts reentering the earth's atmosphere. The lead pair of reindeer will absorb 14.3 QUINTILLION joules of energy per second each. In short, they will burst into flame almost instantaneously, exposing the reindeer behind them, and create deafening sonic booms in their wake. The entire reindeer team will be vaporized within 4.26 thousandths of a second. Santa, meanwhile, will be subjected to centrifugal forces 17,500.06 times greater than gravity. A 250-pound Santa (which seems ludicrously slim) would be pinned to the back of his sleigh by 4,315,015 pounds of force.

In conclusion—If Santa ever DID deliver presents on Christmas Eve, he's dead now!

Source unknown

Love is a matter of chemistry, sex is a matter of physics.

Source unknown

Van Sant, Gus
Like a disc jockey from Paradise, Howard flips Marie over and plays her B side. Every now and then she reaches for Sissy to include her, but the laws of physics insist on being obeyed.

Screenplay of *Even Cowgirls Get the Blues* (p. 34)

von Mises, Richard
The problems of statistical physics are of the greatest interest in our time, since they lead to a revolutionary change in our whole conception of the universe.

Probability, Statistics, and Truth (p. 219)

von Weizsacker, Carl Friedrich
Classical physics has been superseded by quantum theory: quantum theory is verified by experiments. Experiments must be described in terms of classical physics.

Source unknown

Weinberg, Steven
Our job in physics is to see things simply, to understand a great many complicated phenomena, in terms of a few simple principles.

Quoted by Robert K. Adair in
The Great Design (p. 325)

Wells, H.G.
The science of physics is even more tantalizing than it was half a century ago, and, above the level of an elementary introduction, optics, acoustics and the rest, even less teachable. The more brilliant investigators rocket off into mathematical pyrotechnics and return to common speech with statements that are, according to the legitimate meanings of words, nonsensical.

Experiment in Autobiography (p. 176)

Wheeler, John A.
No point is more central than this, that empty space is not empty. It is the seat of the most violent physics.

Quoted by Heinz R. Pagels in
The Cosmic Code (p. 274)

Whitehead, Alfred North
In the present-day reconstruction of physics, fragments of the Newtonian concepts are stubbornly retained. The result is to reduce modern physics to a sort of mystic chant over an unintelligible universe.

Modes of Thought (p. 185)

Physics refers to ether, electrons, molecules, intrinsically incapable of direct observation.

The Principle of Relativity with Application to Physical Science (p. 62)

Wiener, Norbert
Physics—or so it is generally supposed—takes no account of purpose . . .

God and Golem, Inc. (p. 5)

Wilczek, Frank
In physics, you don't have to go around making trouble for yourself—nature does it for you.

Longing for the Harmonies (p. 208)

Witten, Edward
Most people who haven't been trained in physics probably think of what physicists do as a question of incredibly complicated calculations, but that's not really the essence of it. The essence of it is that physics is about concepts, wanting to understand the concepts, the principles by which the world works.

Quoted by Michio Kaku in
Hyperspace (p. 152)

Zolynas, Al
And so, the closer he looks at things, the farther away they seem. At dinner, after a hard day at the universe, he finds himself slipping through his food. His own hands wave at him from beyond a mountain of peas. Stars and planets dance with molecules on his fingertips. After a hard day with the universe, he tumbles through himself, flies through the dream galaxies of his own heart. In the very presence of his family he feels he is descending through an infinite series of Chinese boxes.

The New Physics
The New Physics (p. 55)

PLANETS

Alighieri, Dante
Turning, I perceived
The whiteness round me of the temperate star
The Sixth, whereinto I had been received.

Paradiso
Canto XVII
l. 67–71

Banks, Sir Joseph
Some of our astronomers here incline to the opinion that it is a planet and
not a comet; if you are of that opinion it should forthwith be provided
with a name [or] our nimble neighbors, the French, will certainly save
us the trouble of Baptizing it.

Quoted by Constance A. Lubbock in
The Herschel Chronicle (p. 95)

Borman, Frank
The moon is a different thing to each of us.

From Apollo VIII
December 24, 1968

de Bergerac, Cyrano
I believe the Planets are Worlds about the Sun, and that the fixed Stars are
also Suns, which have Planets about them, that's to say, Worlds which
because of their smallness, and that their borrowed light can-not reach
us, are not discernible by Men in this World . . .

The Comical History of the States and Empires of the World and of the Moon and Sun
The History of the World of the Moon (pp. 13–4)

229

de Chardin, Pierre Teilhard
However inconsiderable they may be in the history of sidereal bodies, however accidental their coming into existence, the planets are finally nothing less than the key points of the Universe. It is through them that the axis of life now passes; it is upon them that the energies of an Evolution principally concerned with the building of large molecules is now concentrated.

Quoted by Roger A. MacGowan and Frederick I. Ordway, III in
Intelligence in the Universe (p. 75)

de Fontenelle, Bernard
. . . you must go a great way to prove that the Earth may be a Planet, the Planets so many Earths, and all the Stars Worlds.

Quoted by Roger A MacGowan and Frederick I Ordway, III in
Intelligence in the Universe (p. 75)

Einstein, Albert
There has been an earth for a little more than a billion years. As for the question of the end of it I advise: Wait and see.

Quoted by Eli Maor in
To Infinity and Beyond: A Cultural History of the Infinite (p. 182)

Feynman, Richard P.
. . . what makes planets go around the sun? At the time of Kepler some people answered this problem by saying that there were angels behind them beating their wings and pushing the planets around in orbit. As you will see, the answer is not very far from the truth. The only difference is that the angels sit in a different direction and their wings push inwards.

The Character of Physical Law
Chapter 1 (p. 18)

What men are poets who can speak of Jupiter if he were like a man, but if he is an immense spinning sphere of methane and ammonia must be silent?

Quoted by Robert Osserman in
Poetry of the Universe (p. 93)

Frost, Robert
The Moon for all her light and grace
Has never learned to know her place.
The notedest astronomers
Have set the dark aside for hers.

The Poetry of Robert Frost
Two Leading Lights
l. 15

Fry, Christopher
. . . the moon is nothing
But a circumambulating aphrodisiac
Divinely subsidized to provoke the world
Into a rising birth-rate.

The Lady's Not for Burning (p. 67)

Herschel, William
It has generally been supposed that it was a lucky accident that brought
this new star to my view; this is an evident mistake. In the regular manner
I examined every star of the heavens, not only of that magnitude but
many far inferior, it was that night *its turn* to be discovered.

Quoted by Constance A. Lubbock in
The Herschel Chronicle (pp. 78–9)

Holmes, Sherlock
"But the Solar System!" I protested.

"What the deuce is it to me?" [Sherlock Holmes] interrupted impatiently:
"You say that we go round the sun. If we went round the moon it would
not make a pennyworth of difference to me . . . "

In Arthur Conan Doyle's
The Complete Sherlock Holmes
A Study in Scarlet

Homer
The thick tresses of gold with which Vulcan had crested the helmet
floated round it, and as the evening star that shines brighter than all
others through the stillness of night, even such was the gleam of the
spear which Achilles poised in his right hand, fraught with the death of
noble Hector.

The Iliad
Book XXII, l. 317

At length as the Morning Star was beginning to herald the light which
saffron-mantled Dawn was soon to suffuse over the sea, the flames fell
and the fire began to die.

The Iliad
Book XXIII, l. 226

Huygens, Christiaan
. . . we may mount from this dull Earth, and viewing it from on high,
consider whether Nature has laid out all her cost and finery upon this
small speck of Dirt.

The Celestial World Discover'd (p. 11)

Longfellow, Henry Wadsworth
There is no light in earth or heaven,
  But the cold light of stars;
And the first watch of night is given
  To the red planet Mars.

The Poetical Works of Henry Wadsworth Longfellow
The Light of Stars

Lovell, James A.
The moon is essentially gray, no color . . . It looks like plaster of Paris,
like dirty beach sand with lots of footprints in it.

Washington Post
Lunar Surface is 'Essentially Gray' like Dirty Sand
Wednesday December 25, 1968
Section A, Column 7

Lucan
Now Mars dominates the Heavens.

Pharsalia
Dramatic Episodes of the Civil Wars
Book One
Figulus the Astrologer, l. 662

Milton, John
To behold the wandering Moon,
Riding near her highest noon,
Like one that had been led astray
Through the heav'n's wide pathless way . . .

Il Penseroso (p. 28)

. . . the Moon, whose Orb
Through Optic Glass the Tuscan artist views
At Ev'ning, from the top of Fesole,
Or in Valdarno, to descry new Lands,
Rivers or Mountains in her spotty Globe.

Paradise Lost
Book I, l. 287–91

Robbins, Tom
Our Moon has surrendered none of its soft charm to technology. The
pitter-patter of little spaceboots has in no way diminished its mystery.

Even Cowgirls Get the Blues
Chapter 19 (p. 60)

Sappho
Stars near the lovely moon cover their own bright faces when she is
roundest and lights up the earth with her silver.

Fragment 4
Part I, 24

Shakespeare, William
The heavens themselves, the planets,
   and this center
Observe degree, priority, and place,
Insisture, course, proportion, season, form,
Office, and custom, in all line of order.

Troilus and Cressida
Act I, scene 3

Shelley, Percy Bysshe
That orbéd maiden
With white fire laden,
   Whom mortals call the Moon

The Complete Poetical Works of Shelley
The Cloud, l. 45–7

Art thou pale for weariness
Of climbing heaven, and gazing on the earth,
Wandering companionless
Among the stars that have a different birth . . . ?

The Complete Poetical Works of Shelley
To the Moon
I

Spenser, Edmund
From hence wee mount aloft into the skie,
And looke into the christall firmament,
There we behold the heavens great hierarchie,
The starres pure light, the spheres swift movement . . .

The Complete Poetical Works of Edmund Spenser
The Teares of the Muses
l. 505–8

Swedenborg, Emanuel
[There are] very many earths, inhabited by man . . . thousands, yea, ten thousands of earths, all full of inhabitants . . . not only in this solar system, but also beyond it, in the starry heaven.

Our Solar System, Whicн are Called Planets, and Earths in the Starry Heavens (1758)
Quoted by Roger A. MacGowan and Frederick I. Ordway, III in
Intelligence in the Universe (p. 1)

Tagore, Rabindranath
Through millions and millions of years,
The stars shine,
Fiery whirlpools revolve and rise
In the dark ever-moving current of time.

In this current
The earth is a bubble of mud . . .

Our Universe (p. 43)

Tennyson, Alfred Lord
Mars
As he glow'd like a ruddy shield on the Lion's breast . . .

Alfred Tennyson's Poetical Works
Maud
Part III, I, l. 13–14

For a breeze of morning moves,
  And the planet of love is on high,
Beginning to faint in the light that she loves
  On a bed of daffodil sky . . .

Alfred Tennyson's Poetical Works
Maud
Part XXII, ii, l. 856–9

This world was once a fluid haze of light,
Till toward the centre set the starry tides,
And eddied into suns, that wheeling cast
The planets.

Alfred Tennyson's Poetical Works
The Princess
Part II, 101

While Saturn whirls, his steadfast shade
Sleeps on his luminous ring.

Alfred Tennyson's Poetical Works
The Palace of Art, l. 15–16

Thurber, James
"The moon is 300,000 miles away," said the Royal Mathematician. "It is round and flat like a coin, only it is made of asbestos, and it is half the size of this kingdom. Furthermore, it is pasted on the sky.

Many Moons

Tolstoy, Alexi
"Which is more useful, the Sun or the Moon?" asks Kuzma Prutkov, the renowned Russian philosopher, and after some reflection he answers himself: "The Moon is the more useful, since it gives us its light during the night, when it is dark, whereas the Sun shines only in the daytime, when it is light anyway."

Quoted by George Gamow in
The Birth and Death of the Sun (p. 1)

Warren, Henry White

Saturn is a world in formative processes. We cannot hear the voice of the Creator there, but we can see matter responsive to the voice, and molded by his word.

Recreations in Astronomy: with directions for practical experiments and telescopic work

Wright, Edward

This Sphaere, is nothing else but a representation of the celestial orbes and circles, that have bene imagined for the easier understanding, expressing, & counting of the motions and appearances, eyther common to the whole heavens, or proper to the Sunne and Moone.

The Description and Use of the Sphaere (p. 1)

Anyone who is not shocked by quantum theory has not understood it.

Niels Bohr – (See p. 244)

PRAYER

Lederman, Leon
Dear Lord, forgive me the sin of arrogance, and Lord, by arrogance I mean the following . . .

Quoted by John D. Barrow in
The Artful Universe (p. 31)

Tukey, John W.
The physical sciences are used to "praying over" their data, examining the same data from a variety of points of view. This process has been very rewarding, and has led to many extremely valuable insights. Without this sort of flexibility, progress in physical science would have been much slower. Flexibility in analysis is often to be had honestly at the price of a willingness not to demand that what has *already* been observed shall establish, or prove, what analysis *suggests*. In physical science generally, the results of praying over the data are thought of as something to be put to further test in another experiment, as indications rather than conclusions.

The Annals of Mathematical Statistics
The Future of Data Analysis (p. 46)
Volume 33, Number 1, March 1962

Unknown
Grant, oh God, Thy benedictions
On my theory's predictions
Lest the facts, when verified,
Show Thy servant to have lied.

Proceedings of the Chemical Society
January 1963 (pp. 8–10)

God grant that no one else has done
  The work I want to do,
Then give me the wit to write it up
  In decent English, too.

Applied Optics
Of Optics and Opticists (p. 273)
Volume 8, Number 2, February 1969

It is impossible to imagine the universe run by a wise, just and omnipotent God, but it is quite easy to imagine it run by a board of gods.

H.L. Mencken – (See p. 104)

PRINCIPLE

Polya, G.
This principle is so perfectly general that no particular application is possible.

How to Solve It (p. 181)

PROBABILITY

Barrow, John
Tipler, Frank
In a randomly infinite Universe, any event occurring here and now with finite probability must be occurring simultaneously at an infinite number of other sites in the Universe. It is hard to evaluate this idea any further, but one thing is certain: if it is true then it is certainly not original!

The Anthropic Cosmological Principle (p. 249)

Borel, Émile
Probabilities must be regarded as analogous to the measurement of physical magnitudes; that is to say, they can never be known exactly, but only within certain approximation.

Probabilities and Life
Introduction (pp. 32–3)

Born, Max
If Gessler had ordered William Tell to shoot a hydrogen atom off his son's head by means of an α particle and had given him the best laboratory instruments in the world instead of a cross-bow, Tell's skill would have availed him nothing. Hit or miss would have been a matter of chance.

Quoted by Sir Arthur Eddington in
New Pathways in Science (p. 82)

Coats, R.H.
. . . the electron is just a "smear of probability".

Journal of the American Statistical Association
Science and Society (p. 6)
Volume 34, Number 205, March 1939

Eddington, Sir Arthur Stanley

But it is necessary to insist more strongly than usual that what I am putting before you is a *model*—the Bohr model atom—because later I shall take you to a profounder level of representation in which the electron, instead of being confined to a particular locality, is distributed in a sort of probability haze all over the atom . . .

New Pathways in Science (p. 34)

In most modern theories of physics probability seems to have replaced aether as "the nominative of the verb 'to undulate' ".

New Pathways in Science (p. 110)

Feynman, Richard P.

A philosopher once said "It is necessary for the very existence of science that the same conditions always produce the same results". Well, they do not.

The Character of Physical Law
Chapter 6 (p. 147)

In its efforts to learn as much as possible about nature, modern physics has found that certain things can never be "known" with certainty. Much of our knowledge must always remain uncertain. The *most* we can know is in terms of probabilities.

The Feynman Lectures on Physics
Volume I
Chapter 6-5 (p. 6-11)

Nature permits us to calculate only probabilities.

QED (p. 19)

Harrison, Edward Robert

We have a wraithlike quantum world of ghostly waves where all is fully determined and predictable. Yet, when we translate it into our observed world of sensible things and their events, we are limited to the concept of chance and the language of probability. What happens at the interface of the quantum world and the observed world may be this or may be that.

Masks of the Universe (p. 124)

Herbert, Nick

probability = (possibility)^2

Quantum Reality (p. 96)

Huygens, Christiaan
. . . I have finally judged that it was better worth while to publish this writing such as it is, than to let it run the risk, by waiting longer, of remaining lost.

There will be seen in it demonstrations of those kinds which do not produce as great a certitude as those of geometry, and which even differ much therefrom, since, whereas the geometers prove their propositions by fixed and incontestable principles, here the principles are verified by the conclusions to be drawn from them; the nature of these things not allowing of this being done otherwise. It is always possible to attain thereby to a degree of probability which very often is scarcely less than complete proof.

Treatise on Light
Preface

Pascal, Blaise
Probability—Each one can employ it; no one can take it away.

Pensées
913

Russell, Bertrand
According to quantum mechanics, it cannot be known what an atom will do in given circumstances; there are a definite set of alternatives open to it, and it chooses sometimes one, sometimes another. We know in what proportion of cases one choice will be made, in what proportion a second, or a third, and so on. But we do not know any law determining the choice in an individual instance. We are in the same position as a booking-office clerk at Paddington, who can discover, if he chooses, what proportion of travelers from that station go to Birmingham, what proportion to Exeter, and so on, but knows nothing of the individual reasons which lead to one choice in one case and another in another.

Religion and Science (pp. 158–9)

Whyte, Lancelot Law
If the universe is a mingling of probability clouds spread through a cosmic eternity of space–time, how is there as much order, persistence and coherent transformation as there is?

Essays on Atomism: from Democritus to 1960
Chapter 2 (p. 27)

PROTONS

Dyson, Freeman J.
The most serious uncertainty affecting the ultimate fate of the universe is the question whether the proton is absolutely stable against decay into lighter particles. If the proton is unstable, all matter is transitory and must dissolve into radiation.

Quoted by G. Borner in
The Early Universe (p. 257)

Eddington, Sir Arthur Stanley
I believe there are 15,747,724,136,275,002,577,605,653,961,181,555,468,044,-717,914,527,116,709,366,231,425,076,185,631,031,296 protons in the universe and the same number of electrons.

The Philosophy of Physical Sciences (p. 170)

Updike, John
Neutrinos make a muon when
a proton, comin' through the rye,
hits a burst of hadrons; then
eureka! γ splits from π.

Tossing and Turning
News from the Underworld

PULSAR

Thomsen, Dietrick E.
Eberhart, Jonathan

Rhythmically pulsating radio source,
Can you not tell us what terrible force
Renders your density all so immense
To account for your signal so sharp and intense?

Are you so dense that no matter you own;
Not atoms nor protons, save neutrons alone?
And do you then fluctuate once every second
So fixed that by you all our clocks might be reckoned?

Or are you two stars bound together in action
That spin like a lighthouse beam gone to distraction?
What in the world can account for your course
O rhythmically pulsating radio source?

And perhaps is there more than your radio beam?
Perhaps visible light in a radiant stream?
An what if the cause of your well-metered twitch
Is a strange but intelligent hand at the switch?

A world of astronomers ponder, a-pacing,
The cause of your infernal, rock-steady spacing,
To see your pulse vary, they valiantly strive,
From 1.337295.

But the biggest of mysteries plaguing our earth
Is, how of your kind can there be such a dearth?
In infinite space one should find ever more
Can it be that your number indeed is but four.

Science News
The Pulsar's Pindar (p. 562)
Volume 93, 15 June 1968

QUANTUM THEORY

Belinfante, Frederik Jozef
If I get the impression that Nature itself makes the decisive choice what possibility to realize, where quantum theory says that more than one outcome is possible, then I am ascribing personality to Nature, that is to something that is always everywhere. Omnipresent eternal personality which is omnipotent in taking the decisions that are left undetermined by physical law is exactly what in the language of religion is called God.

Quoted by John D. Barrow in
The World within the World (p. 157)

Bohr, Niels
Anyone who is not shocked by quantum theory has not understood it.

Quoted by N.C. Panda in
Maya in Physics (p. 73)

Born, Max
[In quantum mechanics] we have the paradoxical situation that observable events obey laws of chance, but that the probability for these events itself spreads according to laws which are in all essential features causal laws.

Natural Philosophy of Cause and Chance (p. 103)

Bridgman, P.W.
The explanatory crisis which now confronts us in relativity and quantum phenomena is but a repetition of what has occurred many times in the past . . ... Every kitten is confronted with such a crisis at the end of nine days.

The Logic of Modern Physics (p. 42)

DeWitt, Bryce
Graham, Neill
No development of modern science has had a more profound impact on human thinking than the advent of quantum theory. Wrenched out

of centuries-old thought patterns, physicists of a generation ago found themselves compelled to embrace a new metaphysics. The distress which this reorientation caused continues to the present day. Basically physicists have suffered a severe loss: their hold on reality.

> Quoted by Nick Herbert in
> *Quantum Reality* (p. 15)

Dirac, Paul Adrien Maurice

The main object of physical science is not the provision of pictures, but in the formulation of laws governing phenomena and the application of these laws to the discovery of new phenomena. If a picture exists, so much the better; but whether a picture exists or not is a matter of only secondary importance. In the case of atomic phenomena no picture can be expected to exist in the usual sense of the word "picture," by which is meant to model functioning essentially on classical lines. One may extend the meaning of the word "picture" to include any way of looking at the fundamental laws which make their self-consistency obvious. With this extension, one may acquire a picture of atomic phenomena by becoming familiar with the laws of quantum theory.

> *The Principles of Quantum Mechanics* (p. 10)

Dyson, Freeman J.

Dick Feynman told me about his "sum over histories" version of quantum mechanics. "The electron does anything it likes," he said. "It goes in any direction at any speed, forward or backward in time, however it likes, and then you add up the amplitudes and it gives you the wave function." I said to him, "You're crazy." But he wasn't.

> In Harry Woolf's (Editor)
> *Some Strangeness in the Proportion* (p. 376)

Eddington, Sir Arthur Stanley

A very useful kind of operator is the selective operator. In my schooldays a foolish riddle was current—"How do you catch lions in the desert?" Answer: "In the desert you have a lot of sand and a few lions; so you take a sieve and sieve out the sand and the lions remain". I recall it because it describes one of the most usual methods used in quantum theory for obtaining anything that we wish to study.

> *New Pathways in Science* (p. 263)

Rather against my better judgment I will try to give a rough impression of the theory. It would probably be wiser to nail up over the door of the new quantum theory a notice, "Structural alterations in progress—No admittance except on business", and particularly to warn the doorkeeper to keep out prying philosophers.

> *The Nature of the Physical World*
> Chapter X
> Development of a New Quantum Theory (p. 211)

Einstein, Albert
The quantum theory gives me a feeling very much like yours. One really ought to be ashamed of its success, because it has been obtained in accordance with the Jesuit maxim: "Let not thy left hand know what thy right hand doeth."

Quoted by Max Born in
The Born–Einstein Letters (p. 11)
Letter of June 4, 1919 to Born

[Quantum theory] If this is correct, it signifies the end of physics as a science.

Quoted by L.I. Ponomarev in
The Quantum Dice (p. 80)

This theory [quantum theory] reminds me a little of the system of delusions of an exceedingly intelligent paranoiac, concocted of incoherent elements of thoughts.

Quoted by Arthur Fine in
The Shaky Game (p. 1)
Letter of July 5, 1952 to D. Lipkin

Quantum mechanics is certainly imposing. But an inner voice tells me that it is not yet the real thing. The theory says a lot, but does not bring us any closer to the secret of the Old One. I, at any rate, am convinced that He does not throw dice.

Quoted by Ronald W. Clark in
Einstein: The Life and Times
Letter to Max Born, 1926

Feynman, Richard P.
... I think I can safely say that nobody understands quantum mechanics. ... Do not keep saying to yourself, if you can possibly avoid it, "But how can it be like that?" because you will get "down the drain", into a blind alley from which nobody has yet escaped. Nobody knows how it can be like that.

The Character of Physical Law
Chapter 6 (p. 129)

It is possible in quantum mechanics to sneak quickly across a region which is illegal energetically.

The Feynman Lectures on Physics
Volume III
Chapter 8-6 (p. 8-12)

... there are certain situations in which the peculiarities of quantum mechanics can come out in a special way on a large scale.

The Feynman Lectures on Physics
Volume III
Chapter 21-1 (p. 21-1)

Harrison, Edward Robert
In the impalpable and seemingly inconsequential entities of the quantum world, one finds the true music and magic of nature.

Masks of the Universe (p. 123)

Heisenberg, Werner
Quantum theory thus provides us with a striking illustration of the fact that we can fully understand a connection though we can only speak of it in images and parables.

Physics and Beyond: Encounters and Conversations (p. 210)

The problem of quantum theory centers on the fact that the particle picture and the wave picture are merely two different aspects of one and the same physical reality.

The Physical Principles of the Quantum Theory (p. 177)

Kramers, Hendrick Anthony
The theory of quanta is similar to other victories in science; for some months you smile at it, and then for years you weep.

Quoted by L.I. Ponomarev in
The Quantum Dice (p. 80)

Kramers, Hendrick Anthony
Gilles, Holst
The theory of quanta can be likened to a medicine that cures the disease but kills the patient.

Quoted by L.I. Ponomarev in
The Quantum Dice (p. 81)

Pagels, Heinz R.
The world changed from having the determinism of a clock to having the contingency of a pinball machine.

The Cosmic Code (p. 13)

Pauli, Wolfgang
I know a great deal. I know too much. I am a quantum ancient.

Quoted by Jeremy Bernstein in
Experiencing Science (p. 102)

Physics is a blind alley again. In any case, it has become too difficult for me, and I would prefer to be a comedian in the cinema, or something like that, and hear no more about physics.

Quoted by L.I. Ponomarev in
The Quantum Dice (p. 81)

Schrödinger, Erwin

If all this damned quantum jumping were really here to stay, I should be sorry I ever got involved with quantum theory.

Quoted by Werner Heisenberg in
Physics and Beyond: Encounters and Conversations (p. 75)

Stapledon, Olaf

... whenever a creature was faced with several possible courses of action, it took them all, thereby creating many distinct temporal dimensions and distinct histories of the cosmos. Since in every evolutionary sequence of the cosmos there were very many creatures, and each was constantly faced with many possible courses, and the combinations of all their courses were innumerable, an infinity of distinct universes exfoliated from every moment of every temporal sequence in this cosmos.

Star Maker
Chapter XV, Section 2
Mature Creating

Unknown

Philosophers have long wondered why socks have this habit of getting lost, and why humans always end up with large collections of unmatched odd socks. One school of thought says that socks are very antisocial creatures, and have a deep sense of rivalry. In particular, two socks of the same design have feelings of loathing towards each other and hence it is nearly impossible to pair them (e.g., a blue sock will usually be found nestling up to a black one, rather than its fellow blue sock).

On the other hand, quantum theorists explain it all by a generalised exclusion principle—it is impossible for two socks to be in the same eigen-state, and when it's in danger of happening, one of the socks has to vanish. Indeed the Uncertainty Principle also comes in—the only time you know where a sock is, is when you're wearing it, and hence unable to be sure exactly how fast it's moving. The moment you stop moving and look at your sock, it then starts falling to pieces, changing colour, or otherwise becoming indeterminate. Either way, socks may possess Colour and Strangeness, but they seem to lack Charm.

Quoted on the Internet

A quantum mechanic's vacation
Had his colleagues in dire consternation.
For while studies had shown
That his speed was well known,
His position was pure speculation.

Source unknown

Wheeler, John A.
So the quantum, fiery creative force of modern physics, has burst forth in eruption after eruption and for all we know the next may be the greatest of all.

> Quoted by Franco Selleri in
> *Quantum Mechanics Versus Local Realism* (p. 47)

There may be no such thing as the "glittering central mechanism of the universe" to be seen behind a glass wall at the end of the trail. Not machinery but magic may be the better description of the treasure that is waiting.

> Quoted by Nick Herbert in
> *Quantum Reality* (p. 29)

Nothing is more important about quantum physics than this: it has destroyed the concept of the world as "sitting out there." The universe afterwards will never be the same.

> Quoted by Jefferson Hane Weaver in
> *The World of Physics*
> Volume II (p. 427)

. . . if one really understood the central point and its necessity in the construction of the world, one ought to be able to state it in one clear, simple sentence. Until we see the quantum principle with this simplicity we can well believe that we do not know the first thing about the universe, about ourselves, and about our place in the universe.

> Quoted by Francesco de Finis (Editor) in
> *Relativity, Quanta and Cosmology in the Development of the Scientific Thought of Albert Einstein*
> Volume II
> The Quantum and the Universe

Yang, Chen N.
To those of us who were educated after light and reason had struck in the final formulation of quantum mechanics, the subtle problems and the adventurous atmosphere of these pre-quantum mechanics days, at once full of promise and despair, seem to take on an almost eerie quality. We could only wonder what it was like when to reach correct conclusions through reasonings that were manifestly inconsistent constituted the art of the profession.

> *Elementary Particles: A Short History of Some Discoveries in Atomic Physics* (p. 9)

QUARK

Melneckuk, Theodore
Poor Gell-Man seeks
But fails to find
The fractioned freaks
He bore in mined.

And yet a Quarck,
Yea, better, three,
Exist in stark
Reality.

The Physics Teacher
The Hunting of the Quarck (p. 415)
Volume 7, Number 7, October 1969

Unknown
O! O! you eight colourful guys
You won't let quarks materialize
You're tricky, but now we realize
You hold together our nucleis.

Quoted by F. Wilczek and B. Devine in
Longing for the Harmonies (p. 200)

Stop the discrimination of quarks on the basis of colour!
Let them all have a voice:
be they red,
be they green,
be they blue,
be they charming or not.

Stop the infrared slavery!

Universal Society for Prevention of Cruelty to Particles
Source unknown

RADAR

Washington Staff
AN ACOUSTIC-GUIDED SUBMUNITION called the BAT may be good against tanks, but not against an F-117. A reader who works on the stealth fighter in Saudi Arabia says bats (the natural ones) occasionally work their way into F-117 hangers [*sic*]. One night a hungry bat turned right into an F-117 rudder and fell stunned to the floor. He flew away groggily, leaving behind a heightened impression of the aircraft's stealth. "I don't know what the radar return is for the vertical tails of the F-117 but I always thought it had to be more than an insect's," the reader said. "I guess I was wrong." There may be some "science" in this—the ultrasound wavelengths used by bats are roughly the same as X-band radar.

Aviation Week and Space Technology
Washington Roundup (p. 17)
Bat Cunning
Volume 135, Number 5, October 14, 1991

RADIATION

Eddington, Sir Arthur Stanley
It has been widely supposed that the ultimate fate of protons and electrons is to annihilate one another, and release the energy of their constitution in the form of radiation. If so it would seem that the universe will finally become a ball of radiation, becoming more and more rarefied and passing into longer and longer wave-lengths. The longest waves of radiation are Hertzian waves of the kind used in broadcasting. About every 1500 million years this ball of radio waves will double its diameter; and it will go on expanding in geometrical progression for ever. Perhaps then I may describe the end of the physical world as—one stupendous broadcast.

New Pathways in Science
Chapter III, VI (p. 71)

Gamow, George
Radiation is like butter, which can be bought or returned to the grocery store only in quarter-pound packages, although the butter as such can exist in any desired amount (not less, though, than one molecule!).

Thirty Years that Shook Physics (pp. 22–3)

Planck, Max
Either the quantum of action was a fictional quantity, then the whole deduction of the radiation law was in the main illusionary and represented nothing more than an empty nonsignificant play on formulae, or the derivation of the radiation law was based on sound physical conception.

Quoted by Jefferson Weaver in
The World of Physics
Volume II (p. 284)

Unknown
My reactor has a first name
It's N - A - V - A - L
My reactor has a second name
It's classified as hell
Oh, I like to scram it everyday
And if you ask me why I'll say,
"Cause radiation has a way
of re-arranging DNA!"

Source unknown

RADIUM

Horton, F.

A radium atom was dying,
  And just ere it burst up for aye,
Corpuscles, which round it were flying,
  These last dying words heard to say . . .
  Oh, I am a radium atom.

The American Physics Teacher
The Radium Atom (p. 181)
Volume 7, Number 3, June 1939

Rickard, Dorothy

Little Willie, full of glee,
Put radium in Grandma's tea.
Now he thinks it quite a lark
To see her shining in the dark.

Little Willie

REALITY

Allen, Woody

I hate reality, but it is still the only place where I can get a decent steak.

Quoted by G. Borner in
The Early Universe (p. 26)

Bohr, Niels

. . . an independent reality in the ordinary physical sense can neither be ascribed to the phenomena nor to the agencies of observation.

Atomic Theory and the Description of Nature
Development of Atomic Theory (p. 54)

Einstein, Albert

Physical concepts are free creations of the human mind, and are not, however it may seem, uniquely determined by the external world. In our endeavor to understand reality we are somewhat like a man trying to understand the mechanism of a closed watch. He sees the face and the moving hands, even hears its ticking but he has no way of opening the case. If he is ingenious he may form some picture of a mechanism which could be responsible for all the things he observes but he may never be quite sure his picture is the only one which could explain his observations. He will never be able to compare his picture with the real mechanism and he cannot even imagine the possibility or the meaning of such a comparison.

The Evolution of Physics (p. 33)

Space has devoured ether and time; it seems to be on the point of swallowing up also the field and the corpuscles, so that it alone remains as the vehicle of reality.

Quoted by R. Thiel in
And There Was Light (p. 345)

All our science, measured against reality, is primitive and childlike—and yet it is the most precious thing we have.

The Physics Teacher
Glimpses of Einstein—a photo essay (p. 200)
Volume 12, Number 4, April 1974

As far as the laws of mathematics refer to reality, they are not certain; and as far as they are certain, they do not refer to reality.

Sidelights on Relativity (p. 28)

Heisenberg, Werner

[The probability wave] meant a tendency for something. It was a quantitative version of the old concept of "Potentia" in Aristotelian philosophy. It introduced something standing in the middle between the idea of an event and the actual event, a strange kind of physical reality just in the middle between possibility and reality.

Physics and Philosophy (p. 41)

Hilbert, David

It has become perfectly clear that physics does not deal with the material world or with the contents of reality, but rather, what it perceives is merely the *formal constitutions* of reality.

Quoted by Walter R. Fuchs in
Mathematics for the Modern Mind (p. 240)

Riordan, Michael

Subatomic reality is a lot like that of a rainbow, whose position is defined only relative to an observer. This is not an objective property of the rainbow-in-itself but involves such subjective elements as the observer's own position. Like the rainbow, a subatomic particle becomes fully "real" only through the process of measurement.

The Hunting of the Quark (p. 39)

Wheeler, John A.

What we call reality consists . . . of a few iron posts of observation between which we fill an elaborate papier-mâché of imagination and theory.

In Harry Woolf's (Editor)
Some Strangeness in the Proportion
Beyond the Black Hole
Chapter 22

RED SHIFT

Boas, Ralph P.
Consider the Pitiful Plight
Of a runner who wasn't too bright,
  But sprinted so fast
  He vanished at last
By red-shifting himself out of sight.

Reprinted in Ralph P. Boas, Jr.'s
Lion Hunting & Other Mathematical Pursuits (p. 103)

Kudlicki, Andrzej
What's the easiest way to observe Doppler's effect optically (not acoustically) in one's everyday life? Go out in the evening and look at the cars. The lights are white or yellow when they approach, but they are red when they are moving away of you.

Quoted on the Internet

RELATIVITY

Buller, Arthur Henry Reginald
There was a young lady named Bright,
Whose speed was far faster than light.
  She set out one day
  In a relative way,
And returned home the previous night.

Punch
Relativity (p. 591)
Volume CLXV, December 19, 1923

Cerf, Bennett
The best of them was the conversation between Ginsberg, who demanded to know what relativity was, and Garfinkle, who brazenly attempted to explain it to him. "It's like this," says Garfinkle. "You go to a dentist to get a tooth pulled. You are in the chair only five minutes, but it hurts so much that you think you are there for an hour. Now on the other hand, you go to see your best girl that same evening. She is in your arms for a full hour, but it is so wonderful to have her there that to you it feels like only five minutes."

Try and Stop Me
Jokes about relativity (p. 163)

Einstein, Albert
When you are courting a nice girl an hour seems like a second. When you sit on a red-hot cinder a second seems like an hour. That's relativity.

News Chronicle
14 March 1949

Of the general theory of relativity you will be convinced, once you have studied it. Therefore, I am not going to defend it with a single word.

Letter to A. Sommerfeld
February 8, 1916

I sometimes ask myself how it came about that I was the one to develop the theory of relativity. The reason, I think, is that a normal adult never stops to think about problems of space and time. These are things which he has thought of as a child. But my intellectual development was retarded, as a result of which I began to wonder about space and time only when I had already grown up.

Quoted by John D. Barrow in
Theories of Everything (p. 68)

Harrison, B.
Thorne, Kip S.
Wakano, M.
Wheeler, John A.
If one intends to abandon Relativity, here is the place to do so. Otherwise one is on the way to a new world of physics, both classical and quantum. Here we go!

Quoted by Jean-Pierre Luminet in
Black Holes (p. 117)

Lawrence, D.H.
I like relativity and quantum theories because I don't understand them and they make me feel as if space shifted about like a swan that can't settle, refusing to sit still and be measured; and as if the atom were an impulsive thing always changing its mind.

Quoted by Leon Lederman in
The God Particle

Mach, Ernst
I can accept the theory of relativity as little as I can accept the existence of atoms and other such dogmas.

Quoted by Stephen Pile in
The Book of Heroic Failures

Nabokov, Vladimir
At this point, I suspect, I should say something about my attitude to "Relativity." It is not sympathetic. What many cosmogonists tend to accept as an objective truth is really the flaw inherent in mathematics which parades as truth.

Ada or Ardor: A Family Chronicle (pp. 577–8)

Rindler, Wolfgang
Relativity has taught us to be wary of time.

Essential Relativity (p. 203)

Rogers, Eric
Since Relativity *is* a piece of mathematics, popular accounts that try to explain it without mathematics are almost certain to fail.

> *Physics for the Inquiring Mind*
> Chapter 31 (p. 472)

Russell, Bertrand
[Einstein's theory of relativity] is probably the greatest synthetic achievement of the human intellect up to the present time.

> *The New York Times*
> Father of Relativity and Outstanding Pacifist Inspired U.S.
> to make Atom Bomb? (p. 24)
> Tuesday April 19, 1955
> Section A, Column 3

Sciama, Dennis
General relativity contains within itself the seeds of its own destruction.

> Quoted by J.D. Barrow in
> *The World Within the World* (p. 306)

Thomson, Sir J.J.
It is not the discovery of an outlying island, but of a whole continent of new scientific ideas of the greatest importance to some of the most fundamental questions connected with physics.

> *The New York Times*
> Eclipse Showed Gravity Variation (p. 6)
> Sunday November 9, 1919
> Section A, Column 3

Unknown
We thought that lines were straight and Euclid true.
God said: "Let Einstein be," and all's askew.

> Quoted by John Boodin in
> *Cosmic Evolution* (p. 295)

There once was a fellow named Fisk,
Whose fencing was exceedingly brisk.
  So fast was his action
  The relativistic contraction
Reduced his rapier to a disc.

> Source unknown

Like two shoes on a shelf, When they questioned her, answered Miss
Bright,
"I was there when I got home that night;
  So I slept with myself,
Put-up relatives shouldn't be tight!"

Quoted by Martin Gardner in
Time Travel (p. 9)

There was a young couple named Bright
Who could make love much faster than light.
  They started one day
  In the relative way,
And came on the previous night.

Quoted by Martin Gardner in
Time Travel and Other Mathematical Bewilderments (p. 10)

Einstein's is a wonderful notion
That a rod will contract when in motion,
All the clocks will go slow,
And yet no one will know!
So the matter need cause no commotion.

The Mathematical Gazette
Volume XI, Number 156, January 1922 (p. 22)

Weyl, Hermann
It is as if a wall which separated us from Truth has collapsed. Wider
expanses and greater depths are now exposed to the searching eye of
knowledge, regions of which we had not even a presentiment. It has
brought us much nearer to grasping the plan that underlies all physical
happening.

Space, Time, and Matter
Preface to the First Edition

Williams, W.H.
You hold that time is badly warped,
That even light is bent:
I think I get the idea there,
If this is what you meant;
The mail the postman brings today,
Tomorrow will be sent.

Quoted by R.W. Clark in
Einstein: The Life and Times
The Einstein and the Eddington (p. 330)

REST

Born, Max

It is odd to think that there is a word for something which, strictly speaking, does not exist, namely, "rest."

The Restless Universe (p. 1)

SIMULTANEITY

Bridgman, P.W.
Einstein, in thus analyzing what is involved in making a judgment of simultaneity, and in seizing on the act of the observer as the essence of the situation, is actually adopting a new point of view as to what the concepts of physics should be, namely, the operational view.

The Logic of Modern Physics
Broad Points of View (p. 8)

Einstein, Albert
The concept does not exist for the physicist until he has the possibility of discovering whether or not it is fulfilled in an actual case . . .. As long as this requirement is not satisfied, I allow myself to be deceived as a physicist (and of course the same applies if I am not a physicist) when I imagine that I am able to attach a meaning to the statement of simultaneity.

Relativity
Chapter VIII (p. 22)

So we see that we cannot attach any *absolute* signification to the concept of simultaneity, but that two events which, viewed from a system of co-ordinates, are simultaneous, can no longer be looked upon as simultaneous events when envisaged from a system which is in motion relative to that system.

The Principles of Relativity (pp. 42–3)

SINGULARITY

Davies, Paul Charles William
When a singularity bursts upon the universe, the rational organization of the cosmos is faced with disintegration.

The Edge of Infinity (p. 129)

SOLAR SYSTEM

Browning, Robert
Ah, but a man's reach should exceed his grasp,
Or what's a heaven for?

The Poems of Robert Browning
Andrea del Sarto
l. 97

Emerson, Ralph Waldo
The solar system has no anxiety about its reputation . . .

The Works of Ralph Waldo Emerson
Volume VI
Conduct of Life
Worship (p. 297)

Jeffrey, Lord
Damn the solar system. Bad light; planets too distant; pestered with comets; feeble contrivance; could make a better myself.

Quoted by John D. Barrow in
The Artful Universe (p. 34)

Patten, W.
A solar system has attributes and powers that can not be defined or measured in terms of its members, or of its ultimate chemical elements, for a solar system is not merely an aggregate, or the algebraic sum of its various elements and qualities . . . .. It is a system, a new type of individuality, with special creative powers of its own.

The Grand Strategy of Evolution (p. 34)

SOUND

Graham, L.A.
Tom, Tom, the piper's son,
Could pipe as pert as any one
In longitudinal waves of strength
With frequency inverse to length.
And every blast, when he'd begin it,
Blared out at thirteen miles a minute.

Ingenious Mathematical Problems and Methods
Mathematical Nursery Rhyme No. 22

Lundberg, Derek
dB or not dB, that is the question.
Whether 'tis number, for if 'tis,
then to it you can add and rightfully sum
    produce,
or, whether 'tis logarithm, for it logarithm
    it be,
then in thinking you add, you are gravely mistaken,
For t'was well known, alas perhaps now
    less so,
if log plus log you add, you doth a product get.

Physics World
dB or not dB? (p. 76)
Volume 6, Number 9, September 1993

Thurber, James
Music is the crystallization of sound.

Journal
Volume I
February 5, 1841

Unknown
A physicist named Haines
After infinite racking of brains
  Now says he has found
  A new kind of sound
That travels much faster than planes.

Source unknown

Somebody has to have the last word. Otherwise, every reason can be met with another one and there would never be no end to it.

Albert Camus – (See p. 122)

SPACE

Alfvén, Hannes
Having probes in space was like having a cataract removed.

Quoted by Eric J. Lerner in
The Big Bang Never Happened (p. 45)

Bailey, Philip James
Unimaginable space,
As full of suns as is earth's sun of atoms.

Festus
IV (p. 61)

Barnes, Bishop
It is fairly certain that our space is finite though unbounded. Infinite space is simply a scandal to human thought.

Quoted by Joseph Silk in
The Big Bang (p. 81)

Bruno, Giordano
There is a single general space, a single vast immensity which we may freely call Void: in it are innumerable globes like this on which we live and grow; this space we declare to be infinite, since neither reason, convenience, sense-perception nor nature assign it a limit.

Quoted by Joseph Silk in
The Big Bang (p. 81)

Eddington, Sir Arthur Stanley
To put the conclusion rather crudely—space is not a lot of points close together; it is a lot of distances interlocked.

The Mathematical Theory of Relativity
Chapter I (p. 10)

Frost, Robert
Space ails us moderns: we are sick with space.
Its contemplation makes us out as small
As a brief epidemic of microbes . . .

The Poetry of Robert Frost
The Lesson for Today
l. 68–70

Gauss, Carl Friedrich
We must confess in all humility that, while number is a product of our mind alone, space has a reality beyond the mind whose rules we cannot completely prescribe.

Quoted by Charles W. Misner *et al* in
Gravitation (p. 195)

Heisenberg, Werner
Space is blue and birds fly through it.

Quoted by Harald Fritzsch in
The Creation of Matter (pp. 12–13)

Hoyle, Fred
Space isn't remote at all. It's only an hour's drive away if your car could go straight upwards.

Observer
September 9, 1979

Jeans, Sir James Hopwood
Space, regarded as a receptacle for radiant energy, is a bottomless pit.

Nature Supplement
November 3, 1928 (p. 698)

Joubert, Joseph
Space is the Stature of God.

Pensées
Number 183

Kant, Immanuel
Space is not a conception which has been derived from outward experiences. For, in order that certain sensations may relate to something without me (that is, to something which occupies a different part of space from that in which I am); in like manner, in order that I may represent them not merely as without, of, and near to each other, but also in separate places, the representation of space must already exist as a foundation. Consequently, the representation of space cannot be borrowed from the relations of external phenomena through experience;

but, on the contrary, this external experience is itself only possible through the said antecedent representation.

The Critique of Pure Reason
First Part
Of Space
Metaphysical Exposition of this Conception, 1

Kirk, Captain James T.
Space, the final frontier . . .

Star Trek (opening mission statement)

Lamb, Charles
Nothing puzzles me more than time and space; and yet nothing troubles me less, as I never think about them.

Quoted by James R. Newman in
The World of Mathematics
Volume One (p. 552)
Letter to Thomas Manning
January 2, 1806

Lewis, Gilbert Newton
. . . when we analyze the highly refined concept of space used by mathematicians we find it to be quite similar to the concept of number.

The Anatomy of Science (p. 29)

Maxwell, James Clerk
. . . the aim of the space-crumplers is to make its curvature uniform everywhere, that is over the whole of space whether that whole is more or less than ∞. The *direction* of the curvature is not related to one of the $x\ y\ z$ more than another or to $-x\ -y\ -z$ so that as far as I understand we are once more on a pathless sea, starless, windless and poleless.

Quoted by John D. Barrow in
The World within the World (p. 107)

Minkowski, Hermann
The views of space and time which I wish to lay before you have sprung from the soil of experimental physics, and therein lies their strength. They are radical. Henceforth space by itself, and time by itself, are doomed to fade away into mere shadows, and only a kind of union of the two will preserve an independent reality.

The Principle of Relativity
Space and Time (p. 75)

Newton, Sir Isaac
Absolute space, in its own nature, without relation to anything external, remains always similar and immovable.

Mathematical Principles of Natural Philosophy
Scholium, II

Poincaré, Henri
Space is only a word that we have believed a thing.

The Foundations of Science
Author's Preface to Translation (p. 5)

Smith, Logan
I think of Space, and the unimportance in its unmeasured vastness, of our toy solar system; I lose myself in speculations of the lapse of Time, reflecting how at the best our human life on this minute and perishing planet is as brief as a dream.

Trivia
Self-Analysis (p. 121)

Wheeler, John A.
The space of quantum geometrodynamics can be compared to a carpet of foam spread over a slowly undulating landscape . . .. The continual microscopic changes in the carpet of foam as new bubbles appear and old ones disappear symbolizes the quantum fluctuations in the geometry.

Battelle Rencontres
Superspace and Quantum Geometrodynamics (p. 264)

Space tells matter how to move . . . and matter tells space how to curve.

Gravitation (p. 23)

Whitehead, Alfred North
All space measurement is from stuff in space to stuff in space.

The Aims of Education (p. 233)

Wieghart, James
The entity, we'll call it S, differed in every way.
While some spun left and some spun right, S would merely stay.
S was neither left nor right nor up nor down, but rather in the middle.
Lacking color and charm and other traits that made its neighbors notable,
S resolved to leave this place and find a spot more suitable.
A quiet place that a colorless, measureless waif would find hospitable.
A spot where an entity without mass, or motion,
would not be likely to cause commotion.
After giving much thought to the matter, and energy too,
S arrived at a solution which it felt would do.
"Empty space is just the place for an orphan entity to spend infinity,"
    S thought.
Alas, although the universe is far and wide, there is no empty space
    inside.
So S went beyond into a black void and found . . . nothing.
"Perfect," it said, "but let there be light."

SPACE
The Orphan Entity

SPACE–TIME

Barnett, Lincoln
. . . the universe is not a rigid and inimitable edifice where independent matter is housed in independent space and time; it is an amorphous continuum, without any fixed architecture, plastic and variable, constantly subject to change and distortion. Wherever there is matter and motion, the continuum is disturbed. Just as a fish swimming in the sea agitates the water around it, so a star, a comet, or a galaxy distorts the geometry of the space–time through which it moves.

The Universe and Dr. Einstein (pp. 81–2)

Eddington, Sir Arthur Stanley
It is a *thing*; not like space, which is a mere negation; nor like time, which is—Heaven knows what!

The Nature of the Physical World
Introduction (p. ix)

Maxwell, James Clerk
March on, symbolic host! with step sublime,
Up to the flaming bounds of Space and Time!
There pause, until by Dickenson depicted,
In two dimensions, we the form may trace
Of him whose soul, too large for vulgar space,
In *n* dimensions flourished unrestricted.

In Lewis Campbell and William Garnett's
The Life of James Clerk Maxwell
To the Committee of the Cayley Portrait Fund (p. 637)

Minkowski, Hermann
From this hour on, space as such and time as such shall recede to the shadows and only a kind of union of the two retain significance.

Einstein: A Centenary Volume (p. 231)

Murchie, Guy
... the key to comprehending space–time is the obvious (to me) fact that space is the relationship between things and other things while time is the relationship between things and themselves.

The Seven Mysteries of Life (p. 331)

Pope, Alexander
Ye Gods! annihilate but space and time,
And make two lovers happy.

The Art of Sinking in Poetry
Chapter 11

Reichenbach, Hans
It appears that the solution of the problem of time and space is reserved to philosophers who, like Leibnitz, are mathematicians, or to mathematicians who, like Einstein, are philosophers.

Quoted by Paul Schlipp in
Albert Einstein: Philosopher-Scientist
The Philosophical Significance of the Theory of Relativity, IV

Synge, J.L.
Anyone who studies relativity without understanding how to use simple space–time diagrams is as much inhibited as a student of functions of a complex variable who does not understand the Argand diagram.

Relativity: The Special Theory (p. 63)

Wells, H.G.
To a certain point it had all been plain sailing, a pretty science, with pretty sub-divisions, optics, acoustics, electricity and magnetism and so on. Up to that point, the time-honoured terms which have crystallized out in language about space, speed, force and so forth sufficed to carry what I was learning. All went well in the customary space–time framework. Then things became difficult.

Experiment in Autobiography (pp. 175–6)

Wheeler, John A.
Venture far
To see the nearby
With new eyes.
Perceive yesterday's gravity,
Whether acting on man or mass,
As today's free float.
In the movement of the mass
Grasp the message of the medium:
"I, medium that grips you,

Man or mass,
And tells you how to move
Am not space.
I am spacetime."

A Journey into Gravity and Spacetime
Chapter 2 (p. 17)

Oh event,
Sparkling grain of sand
On the fabric of existence,
Oh interval,
Gossamer tie
Between event and event,
You two tear away the clouds
Of "absolute space" and "absolute time"
And reveal to us spacetime—
Spacetime as doorway,
Doorway, daring traveler,
To the enormity
Of space and time
Open to our visitation.

A Journey into Gravity and Spacetime
Chapter 3 (p. 35)

SPECTRAL SEQUENCE

Unknown
Oh Be a Fine Girl, Kiss Me

Source unknown

Oven Baked Ants, Fried Gently, Kept Moist, Retain Natural Succulence.

Source unknown

Our Best Aid for Gnat Killing—My Raindeer's Nose. Snort.

Source unknown

Owen Gingerich's astronomy course at Harvard University
Oh bring another fully grown kangaroo, my recipe needs some.

Hendry, Khati

Oh brutal and fearsome gorilla, kill my roommate next Saturday.

Luxenberg, Steven M.

Obese but alluring fertility goddess, keep my Radcliffe nymph sensuous.

Berkman, Jim

On bad afternoons, fermented grapes keep Mrs. Richard Nixon smiling.

Metzner, Richard H.

Ornery Bostonians angrily fight Governor King's measures regarding needless sobriety.

Rusk, Michelle

Out by a foot? God knows most referees need sight!

Steyn, Lawrence

Oh Brooke's a famous girl—Klein makes really narrow slacks.

Ellertson, Lottie

Out beyond Andromeda, fiery gases kindle many red new stars.

Elliot, Steven H.

Oh backward astronomer, forget geocentricity! Kepler's motions reveal nature's simplicity.

Gorman, Mike
Baker, Brad

Organs blaring and fugues galore, Kepler's music reads nature's score.

Gomez, John
Mercury
OBAFGKMRNS (p. 38)
Volume 24, Number 2, March–April 1995

Yervant Terzian at Cornell University

Oh bright aurora, flame gently, kindle my radiant night skies.

Our best answer for government: Karl Marx's radically new system.

Oranges, bananas, apples, figs, grapes, kiwis, melons, raspberries, nectarines, strawberries.

On burgers and franks, get ketchup, mustard, relish; not syrup.

One bright afternoon, five grubby kids maliciously ruined Nora's sidewalk.

Offensive backs and field-goal kickers make ridiculously notable salaries.

Out beyond amber fields, graceful knolls must roll, never subsiding.

"Only bright apples fall groundward knocking me," remarked Newton solemnly.

Outdistanced by a few goddamn kilometers, man reaches no star.

One boy and fifty girls kissed madly, rumors now say!

Mercury
OBAFGKMRNS (p. 38)
Volume 24, Number 2, March–April 1995

SPECTROSCOPE

Huggins, Sir William
One important object of this original spectroscopic investigation of the light of the stars and other celestial bodies, namely to discover whether the same chemical elements as those of our earth are present throughout the universe, was most satisfactorily settled in the affirmative; a common chemistry, it was shown, exists throughout the universe.

The Scientific Papers of Sir William Huggins
Spectra of the Fixed Stars (p. 49)

STARS

Aeschylus

I know the nightly concourse of the stars . . .

Agamemnon
l. 4

Aratus

. . . In his fell jaw
Flames a star above all others with searing beams
Fiercely burning, called by mortals Sirius.

Quoted by Garrett P. Serviss in
Astronomy with the Naked Eye (p. 42)

Aurelius, Marcus [Antoninus]

The Pythagoreans bid us in the morning look to the heavens that we may be reminded of those bodies which continually do the same thing and in the same manner perform their work, and also be reminded of their purity and nudity. For there is no veil over a star.

The Meditations of Marcus Aurelius
Book XI, paragraph 27

Brecht, Bertolt

SAGREDO (hesitates to go to the telescope): I feel something almost like fear, Galilei.

GALILEI: Now I shall show you one of milky-white luminous clouds in the Milky Way. Tell me what it consists of!

SAGREDO: Those are stars, innumerable stars.

The Life of Galilie
Quoted by Rudolf Kippenhahn in
Light from the Depths of Time (p. 6)

Brood, William J.
A telescope in the void recently found cosmic "maternity wards" where clouds of interstellar gas and dust appear to be in various stages of giving birth to stars.

The New York Times
'Golden Age' of Astronomy Peers to the Edge of the Universe (p. 1)
Tuesday May 8, 1984
Section C

Browning, Robert
All that I know
  Of a certain star,
Is, it can throw,
  (Like the angled spar)
Now a dart of red,
  Now a dart of blue.

Poems of Robert Browning
Dramatic Lyrics
My Star

Bunting, Basil
Furthest, fairest thing, stars,
  free of our humbug,
each his own, the longer known, the
  more alone,
wrapt in emphatic fire roaring out
  to a black flue . . .

The Complete Poems
Briggflats
Part V

Byron, Lord George Gordon
Ye stars! which are the poetry of heaven!

Childe Harold's Pilgrimage
Canto III, lxxxviii

Clarke, Arthur C.
Overhead, without any fuss, the stars were going out.

The Nine Billion Names of God (p. 11)

Copernicus, Nicolaus
. . . the first and highest of all is the sphere of the fixed stars, which comprehends itself and all things, and is accordingly immovable.

On the Revolutions of the Heavenly Spheres
Book I
On the Order of the Celestial Orbital Circles

Davies, Paul Charles William
Many billions of years will elapse before the smallest, youngest stars complete their nuclear burning and shrink into white dwarfs. But with slow, agonizing finality perpetual night will surely fall.

The Last Three Minutes (p. 50)

Dick, Thomas
Come forth, O man! yon azure round survey,
And view those lamps which yield eternal day.
Bring forth thy glasses; clear thy wondering eyes;
Millions beyond the former millions rise;
Look further;—millions more blaze from yonder skies.

The Works of Thomas Dicks, LL.D.
The Solar System
Volume X
Chapter VIII (p. 197)

Dickinson, Emily
"Arcturus" is his other name—
I'd rather call him "Star."
Its very mean of Science
To go and interfere!

Poems of Emily Dickinson
Old Fashioned

Disraeli, Benjamin
It shows you exactly how a star is formed; nothing can be so pretty! A cluster of vapor, the cream of the milky way, a sort of celestial cheese, churned into light . . .

Tancred (p. 112)

Eddington, Sir Arthur Stanley
The inside of a star is a hurly-burly of atoms, electrons and aether waves.

The Internal Constitution of Stars (p. 19)

Eliot, George
Stars are the golden fruit of a tree beyond reach.

Quoted by Jean-Pierre Luminet in
Black Holes (p. 59)

Emerson, Ralph Waldo
If the stars should appear one night in a thousand years, how would men believe and adore and preserve for many generations the remembrance of the city of God which had been shown.

The Works of Ralph Waldo Emerson
Volume I
Nature (p. 7)

But every night comes out the envoys of beauty, and light the universe with their admonishing smile.

The Works of Ralph Waldo Emerson
Volume I
Nature (p. 7)

Feynman, Richard P.
Poets say science takes away from the beauty of the stars—mere globs of gas atoms. Nothing is "mere." I too can see the stars on a desert night, and feel them. But do I see less or more? The vastness of the heavens stretches my imagination—stuck on this carousel, my little eye can catch one-million-year-old light.

The Feynman Lectures on Physics
Volume I
Chapter 3-4 (footnote, p. 3-6)

Fraunhofer, Joseph von
Approximavit sidera
[He brought the stars closer]

Epitaph on his gravestone

Frost, Robert
They cannot scare me with their empty spaces
Between stars—on stars where no human race is.

The Poetry of Robert Frost
Desert Places

Gamow, George
Twinkle, twinkle, quasi-star
Biggest puzzle from afar
How unlike the other ones
Brighter than a billion suns
Twinkle, twinkle, quasi-star
How I wonder what you are.

Quoted by Louis Berman in
Exploring the Cosmos (p. 311)

Whereas all humans have approximately the same life expectancy the life expectancy of stars varies as much as from that of a butterfly to that of an elephant.

A Star Called the Sun (p. 145)

Goethe, Johann Wolfgang von
Stars, you are unfortunate, I pity you,
Beautiful as you are, shining in your glory . . .

Selected Poems
Night Thoughts

Hardy, Thomas
The sovereign brilliancy of Sirius pierced the eye with a steely glitter . . .

Far from the Madding Crowd
Chapter Two (p. 32)

Hegel, Georg Wilhelm Friedrich
The stars are not pulled this way and that by mechanical forces; theirs is a free motion. They go on their way, as the ancients said, like the blessed gods.

Werke
Bd. 7, Abt. I, p. 97

Hein, Robert
The stars are luminous dandruff from the deity's beard.
When the god-head combs his hair
A new star appears in the sky;
Yet God is not almost bald . . .

Quest of the Singing Tree
Stanzas on the Stars

Herschel, Sir John
Stargazing Knight Errant, beware of the day
When the Hottentots catch thee observing away!
Be sure they will pluck thy eyes out of their sockets
To prevent thee from stuffing stars in thy pockets.

Herschel at the Cape (p. 80)

Hopkins, Gerard Manley
Look at the stars! look, look up at the skies!
O look at all the fire-folk sitting in the air!

The Poetical Works of Gerard Manley Hopkins
The Starlight Night

Huygens, Christiaan
For if 25 years are required for a Bullet out of a Cannon, with its utmost Swiftness, to travel from the Sun to us . . . such a Bullet would spend almost seven hundred thousand years in its Journey between us and the fix'd Stars. And yet when in a clear night we look upon them, we cannot think them above some few miles over our heads.

The Celestial World Discover'd (pp. 154–5)

Jeans, Sir James Hopwood
Any small bit of the sky does not look very different from what it would if bright and faint stars had been sprinkled out of a celestial pepperpot.

The Universe Around Us (p. 37)

Jennings, Elizabeth
The radiance of the star that leans on me
Was shining years ago. The light that now
Glitters up there my eyes may never see,
And so the time lag teases me . . .

Selected Poems
Delay

Longfellow, Henry Wadsworth
Silently one by one, in the infinite meadows of heaven
Blossomed the lovely stars, the forget-me-nots of the angels.

The Poetical Works of Henry Wadsworth Longfellow
Evangeline
Part iii

Milton, John
The stars,
That nature hung in heaven, and filled their lamps
with everlasting oil, to give due light
To the misled and lonely traveler.

Comus
l. 197–200

The stars, she whispers, blindly run;
    A web is wov'n across the sky . . .

In Memoriam
Third section

Newton, Sir Isaac
. . . what hinders the fixed stars from falling upon one another?

Opticks
Book III, Part I, Query 28

Noyes, Alfred
And all those glimmerings where the abyss
    of space
Is powdered with a milky dust, each grain
A burning sun, and every sun the lord
Of its own darkling planets— . . .

Watchers of the Sky
Prologue (p. 8)

Could new stars be born?
Night after night he watched that miracle
Growing and changing colour as it grew . . .

Watchers of the Sky
Tycho Brahe (p. 57)

O'Casey, Sean
. . . an me lookin' up at the sky an' sayin' "what is the stars, what is the stars?"

Juno and the Paycock
Act I (p. 25)

Old Woman
The stars I know and recognize and even call by name. They are my names, of course; I don't know what others call the stars. Perhaps I should ask the priest. Perhaps the stars are God's to name, not ours to treat like pets . . .

Quoted by Robert Coles in
The Old Ones of New Mexico
Two Languages, One Soul (p. 10)

Pagels, Heinz R.
Stars are like animals in the wild. We may see the young but never their actual birth, which is a veiled and secret event.

Perfect Symmetry
Chapter 2 (p. 44)

Pasachoff, Jay M.
Twinkle, twinkle, pulsing star
Newest puzzle from afar.
Beeping on and on you sing,
Are you saying anything?
Twinkle, twinkle more, pulsar,
How I wonder what you are.

Physics Today
Pulsars in Poetry (p. 19)
Volume 22, Number 2, February 1969

Plato
. . . when all the stars which were necessary to the creation of time had attained motion suitable to them, and had become living creatures having bodies fastened by vital chains, and learnt their appointed task . . .

Timaeus
Section 38

Poe, Edgar Allan
Were the succession of stars endless, then the background of the sky would present us a uniform luminosity, like that displayed by the galaxy—since there could be absolutely no point, in all that background, at which would not exist a star. The only mode, therefore, in which, under such a state of affairs, we could comprehend the voids which our telescopes find in innumerable directions, would be by supposing the

distance of the invisible background so immense that no ray from it has yet been able to reach us at all.

Eureka (p. 273)

Poincaré, Henri
The stars are majestic laboratories, gigantic crucibles, such as no chemist could dream.

The Foundations of Science
The Value of Science
Astronomy (p. 295)

Rilke, Rainer Maria
. . . between stars, what distances . . .

Sonnets to Orpheus
Second Part
XX

Russell, Peter
. . . The fixed stars
Are moving really, and the whole Galaxy turning
Round and round on its own axis agitatedly . . .

All for the Wolves
Elegiac

Seneca
There is no easy way to the stars from the earth.

Hercules Furens
Act II

Smith, Logan
"But what are they really? What do they say they are?" the young lady asked me. We were looking up at the Stars . . .

Trivia
The Starry Heaven (p. 51)

Stapledon, Olaf
Very soon the heavens presented an extraordinary appearance, for all the stars directly behind me were now deep red, while those directly ahead were violet. Rubies lay behind me, amethysts ahead of me. Surrounding the ruby constellations there spread an area of topaz stars, and round the amethyst constellations an area of sapphires.

Star Maker
Chapter II

Taylor, Anne
Twinkle, twinkle, little star!

Rhymes for the Nursery
The Star

Tennyson, Alfred Lord
Many a night I saw the Pleiads, rising
  thro' the mellow shade,
Glitter like a swarm of fireflies, tangled in
  a silver braid.

Poems of Tennyson
Locksley Hall

"The stars," she whispers, "blindly run:
A web is wov'n across the sky;
From our waste places comes a cry,
And murmurs from the dying sun."

Poems of Tennyson
In Memoriam
iii

Thoreau, Henry David
When I look at the stars, nothing which the astronomers have said
attaches to them, they are so simple and remote.

Journal
The Red of the Young Oak
Volume VII, September 25, 1854

Twain, Mark
We had the sky, up there, all speckled with stars, and we used to lay
on our backs and look up at them, and discuss about whether they was
made, or only just happened.

Huckleberry Finn
Chapter XIX

Unknown
Twinkle, twinkle, little star
I don't wonder what you are,
For by the spectroscopic ken
I know that you are hydrogen.

Quoted by Douglas Bush in
Science and English Poetry (p. 143)

Twinkle, twinkle little star
How I wonder where you are.
"1.75 seconds of arc from where I seem to be
For $ds^2 \cong (1 - 2GM/r)\, dt^2 - r^2\, d\theta^2 - r^2 \sin^2 \theta\, d\phi^2$."

Quoted by Hans C. Ohanian in
Gravitation and Spacetime (p. 87)

Updike, John
Welcome, welcome, little star
I'm delighted that you are
Up in Heaven's vast extent,
No bigger than a continent.

Telephone Poles and Other Poems
White Dwarf

When, on those anvils at the center of stars,
    and those event more furious anvils
        of the exploding supernovae,
    the heavy elements were beaten together
        to the atomic number 94 . . .

Facing Nature
Ode to Crystallization

Whitman, Walt
I believe a leaf of grass is no less than the journey-work of the stars.

Complete Poems and Collected Prose
Volume 1
Song of Myself, stanza 31

Wilcox, E.W.
Since Sirius crossed the Milky Way
    Full sixty thousand years have gone,
Yet hour by hour and day by day
    This tireless star speeds on and on.

Quoted by Garrett P. Serviss in
Astronomy with the Naked Eye (p. 43)

Wilde, Oscar
LORD DARLINGTON: We are all in the gutter, but some of us are
looking at the stars.

Lady Windermere's Fan
Act Three

Williams, Sarah
I have loved the stars too fondly to be fearful of the night.

The Best Loved Poems of the American People
The Old Astronomer to His Pupil

Wolf, Fred Alan
Stars, like little lost children seeking shelter on a cold night, tend to
cluster, via gravitationally induced starlight, into galaxies.

Parallel Universes (p. 71)

Wordsworth, William
The stars are mansions built by Nature's hand,
And, haply, there the spirits of the blest
Dwell clothed in radiance, their immortal vest;

The Poetical Works of William Wordsworth
Miscellaneous Sonnets XXV

Yeats, William Butler
Under the passing stars, foam of the sky
Live on this lonely face.

The Collected Poems of W.B. Yeats
The Rose of the World

Young, Edward
How distant some of these nocturnal Suns?
So distant (says the Sage) 'twere not absurd
To doubt, if Beams, set out at *Nature's* Birth,
Are yet arriv'd at this so foreign World;

Night Thoughts
Night IX, l. 1226–9

STATISTICS

Eddington, Sir Arthur Stanley
We must not think about space and time in connection with an individual quantum; and the extension of a quantum in space has not real meaning. To apply these conceptions to a single quantum is like reading the Riot Act to one man. A single quantum has not traveled 50 billion miles from Sirius; it has not been 8 years on the way. But when enough quanta are gathered to form a quorum there will be found among them statistical properties which are the genesis of the 50 billion miles' distance of Sirius and the 8 years' journey of the light.

The Nature of the Physical World
Chapter X (pp. 200–1)

Edgeworth, Francis Ysidro
In short, Statistics reigns and revels in the very heart of Physics.

Journal of the Royal Statistical Society
On the Use of the Theory of Probabilities in Statistics Relating to Society (p. 167)
January 1913

Maxwell, James Clerk
. . . molecular science teaches us that our experiments can never give us anything more than statistical information, and that no law deduced from them can pretend to absolute precision. But when we pass from the contemplations of our experiment to that of the molecules themselves, we leave the world of chance and change, and enter a region where everything is certain and immutable.

The Scientific Papers of James Clerk Maxwell
Molecules (p. 374)

Russell, Bertrand
I come now to the statistical part of physics, which is concerned with the study of large aggregates. Large aggregates behave almost exactly as they were supposed to do before quantum theory was invented, so that in regard to them the older physics is very nearly right. There is,

however, one supremely important law which is only statistical; this is the second law of thermodynamics. It states, roughly speaking, that the world is growing continuously more disorderly.

The Scientific Outlook
Scientific Metaphysics (p. 92)

Toulmin, S.
Goodfield, J.
The general public today is so inured to astronomical statistics that people will say unthinkingly, "The stars are an incredible distance away"—and leave matters at that. Greek astronomers were more particular about their arguments, and declined to believe the unbelievable.

The Fabric of the Heavens (p. 124)

von Mises, Richard
The problems of statistical physics are of the greatest in our time, since they lead to a revolutionary change in our whole conception of the universe.

Probability, Statistics, and Truth (p. 219)

Weyl, Hermann
Thus we do not know and cannot know what the individual photon or electron does under given conditions; we can only, with the uncertainty adhering to statistics, predict from the wave image their average behavior under the same conditions.

Mind and Nature (p. 89)

Wigner, Eugene
With classical thermodynamics, one can calculate almost everything crudely; with kinetic theory, one can calculate fewer things, but more accurately; and with statistical mechanics, one can calculate almost nothing exactly.

In Edward B. Stuart, Benjamin Gal-Or and Alan J. Brainard (Editors)
A Critical Review of Thermodynamics (p. 205)

SUN

Adams, Douglas
Several billion trillion tons of superhot exploding hydrogen nuclei rose slowly above the horizon and managed to look small, cold and slightly damp.

Life, the Universe and Everything
Chapter 7

Allen, Woody
The sun, which is made of gas, can explode at any moment, sending our entire planetary system hurtling to destruction; students are advised what the average citizen can do in such a case.

Getting Even (p. 58)

Copernicus, Nicolaus
In the center of all rests the sun. For who would place this lamp of a very beautiful temple in another or better place than this wherefrom it can illuminate everything at the same time.

On the Revolutions of the Heavenly Spheres
Book I
On the Order of the Celestial Orbital Circles

Dryden, John
And since the vernal equinox, the sun
In Aries twelve degrees, or more, had run . . .

The Poetical Works of Dryden
The Cock and the Fox
l. 448–9

Joshua 10:12–13
Then spake Joshua to the Lord in the day when the Lord delivered up the Amorites before the children of Israel, and he said in the sight of Israel, Sun, stand thou still upon Gibeon; and thou, Moon, in the valley of Ajalon.

And the sun stood still, and the moon stayed, until the people had avenged themselves upon their enemies. Is not this written in the book of Jasher? So the sun stood still in the midst of heaven, and hasted not to go down the whole day.

The Bible

Mann, Thomas
"He does seem rather weird," was Hans Castorp's view. "Some of the things he said were very queer: it sounded as if he meant to say that the sun revolves round the earth."

The Magic Mountain
Of City of God, and Deliverance by Evil (p. 407)

Mayer, Robert
The Sun . . . is an inexhaustible source of physical force—that continuously wound-up spring which sustains in motion the mechanism of all the activities on Earth.

Quoted by L.I. Ponomarev in
The Quantum Dice (p. 228)

Plato
That there might be some visible measure of their relative swiftness and slowness as they proceeded in their eight courses, God lighted a fire, which we now call the sun, in the second from the earth of these orbits . . .

Timaeus
Section 39

Raymo, Chet
For 5 billion years the sun has exhaled a faint breath as it burns, bathing the Earth in the flux of its exhalations, a wind of atoms and subatomic particles that feeds the Earth's atmosphere and ignites auroras.

The Soul of the Night (p. 81)

Rutherford, Mark [William Hale White]
The sun, we say, is the cause of heat, but the heat *is* the sun, hence on this window-ledge.

More Pages from a Journal

Smart, Christopher
Glorious the sun in mid-career;
Glorious th' assembled fires appear.

Poems by Christopher Smart
A Song to David
LXXXIV

Updike, John
The zeros stared back, every one a wound leaking the word "poison."
"That's the weight of the Sun," Caldwell said.

The Centaur (p. 37)

The only distinction between physicists and engineers
is the physicists have more questions than answers
while engineers have more answers than questions.

Unknown – (See p. 204)

SYMBOL

Eddington, Sir Arthur Stanley
The symbol A is not the counterpart of anything in familiar life. To the child the letter A would seem horribly abstract; so we give him a familiar conception along with it. "A was an Archer who shot at a frog." This tides over his immediate difficulty; but he cannot make serious progress with word-building as long as Archers, Butchers, Captains, dance round the letters. The letters are abstract, and sooner or later he has to realise it. In physics we have outgrown archer and apple-pie definitions of the fundamental symbols. To a request to explain what an electron really is supposed to be we can only answer, "It is part of the A B C of physics".

The Nature of the Physical World
Introduction (p. xiv)

SYMMETRY

Aristotle
A nose which varies from the ideal of straightness to a hook or snub
may still be of good shape and agreeable to the eye.

Politics
Book V, Chapter 9, 1309b, [20]

Blake, William
Tyger, Tyger, burning bright
In the forest of the night,
What immortal hand or eye
Could frame thy fearful symmetry?

The Complete Writings of William Blake
Songs of Innocence and of Experience
The Tyger

Borges, Jorge Luis
Reality favors symmetry.

Conversations with Jorge Luis Borges
Richard Durgin (Editor)

Carroll, Lewis
You boil it in sawdust: you salt it in glue:
You condense it with locust and tape:
Still keeping one principle object in view—
To preserve its symmetrical shape.

The Complete Writings of Lewis Carroll
The Hunting of the Snark
Fit the Fifth
The Beaver's Lesson

Feynman, Richard P.

Why is nature so nearly symmetrical? No one has any idea why. The only thing we might suggest is something like this: There is a gate in Japan, a gate in Neiko, which is sometimes called by the Japanese the most beautiful gate in all Japan; it was built in a time when there was great influence from Chinese art. The gate is very elaborate, with lots of gables and beautiful carvings and lots of columns and dragon heads and princes carved into the pillars, and so on. But when one looks closely he sees that in the elaborate and complex design along one of the pillars, one of the small design elements is carved upside down; otherwise the thing is completely symmetrical. If one asks why this is, the story is that it was carved upside down so that the gods will not be jealous of the perfection of man. So they purposely put the error in there, so that the gods would not be jealous and get angry with human beings.

We might like to turn the idea around and think that the true explanation of the near symmetry of nature is this: that God made the laws only nearly symmetrical so that we should not be jealous of His perfection!

The Feynman Lectures on Physics
Symmetry in Physical Law
Volume I, Chapter 52, Section 5-9 (p. 52-12)

Herbert, George

My body is all symmetry,
Full of proportions, one limb to another,
And all to all the world besides:
Each part may call the farthest, brother:
For head with foot hath private smity,
And both with moon and tides.

The Works of George Herbert
The Temple
Man

Mackay, Charles

Truth . . . and if mine eyes
Can bear its blaze, and trace its symmetries,
Measure its distance, and its advent wait,
I am no prophet—I but calculate.

The Poetical Works of Charles Mackay
The Prospects of the Future

Mao Tse-tung

Tell me why should symmetry be of importance?

30 May, 1974

Pascal, Blaise

Those who make antitheses by forcing words are like those who make false windows for symmetry.

Pensées
Section I, 27

Symmetry is what we see at a glance . . .

Pensées
Section I, 28

Updike, John

When you look	kool uoy nehW
into a mirror	rorrim a otni
it is not	ton si ti
yourself you see,	,ees uoy flesruoy
but a kind	dnik a tub
of apish error	rorre hsipa fo
posed in fearful	lufraef ni desop
symmetry	yrtemmys

Telephone Poles and Other Poems
Mirror

Valéry, Paul

The universe is built on a plan the profound symmetry of which is somehow present in the inner structure of our intellect.

Quoted by Jefferson Hane Weaver in
The World of Physics
Volume II (p. 521)

Weyl, Hermann

Symmetry, as wide or as narrow as you may define its meaning, is one idea by which man through the ages has tried to comprehend and create order, beauty, and perfection.

Symmetry (p. 5)

Symmetry is a vast subject, significant in art and nature. Mathematics lies at its root, and it would be hard to find a better one on which to demonstrate the working of the mathematical intellect.

Quoted by Stanley Gudder in
A Mathematical Journey (p. 292)

Yang, Chen N.

Nature seems to take advantage of the simple mathematical representations of the symmetry laws. When one pauses to consider the elegance and the beautiful perfection of the mathematical reasoning involved and contrast it with the complex and far-reaching physical consequences, a deep sense of respect for the power of the symmetry laws never fails to develop.

Quoted by Heinz R. Pagels in
The Cosmic Code (p. 289)

Zee, Anthony

Pick your favorite group: write down the Yang–Mills theory with your groups as its local symmetry group; assign quark fields, lepton fields, and Higgs fields to suitable representations; let the symmetry be broken spontaneously. Now watch to see what the symmetry breaks down to ... that, essentially, is all there is to it. Anyone can play. To win, one merely has to hit on the choice used by the Greatest Player of all time. The prize? Fame and glory, plus a trip to Stockholm.

Fearful Symmetry (pp. 253–4)

"Just keep away from the black hole garbage bin by the door as you leave."

Malcolm Longair – (See p. 35)

SYSTEMS

Coates, Robert
He has so clearly laid open and set before our eyes the most beautiful frame of the System of the World, that if King Alphonse were now alive he would not complain for want of the graces of simplicity or of harmony in it.

Quoted by Robert H. March in
Physics for the Poets
Preface to the Principia

TACHYONS

Herbert, Nick

Although most physicists today place the probability of the existence of tachyons only slightly higher than the existence of unicorns, research into the properties of these hypothetical FTL [faster than light] particles has not been entirely fruitless.

Faster than Light (p. 137)

TELESCOPE

Beggs, James
Boland, Edward P.
James Beggs: The Space Telescope will be the eighth wonder of the world.

Edward P. Boland: It ought to be at that price.

Quoted by Robert Smith in
The Space Telescope (Prior to Table of Contents)

Bierce, Ambrose
TELESCOPE, *n*. A device having a relation to the eye similar to that of the telephone to the ear, enabling distant objects to plague us with a multitude of needless details. Luckily it is unprovided with a bell summoning us to the sacrifice.

The Enlarged Devil's Dictionary

Brecht, Bertolt
I saw a brand-new instrument in Amsterdam. A tube affair. "See things five times as large as life!" It had two lenses, one at each end, one lens bulged and the other was like that.

Galileo
Scene One

Cedering, Siv
I have helped him polish the mirrors
and lenses of our new telescope. It is
the largest in existence. Can you imagine
the thrill of turning it to some new
corner of the heavens to see
something never before seen
from the earth? . . .

Letters from the Floating World
Letter from Caroline Herschel (1750–1840) (p. 116)

Durant, Will
. . . for lack of a telescope Aristotle's astronomy is a tissue of childish romance.

The Story of Philosophy (p. 64)

Frost, Robert
He burned his house down for the fire insurance
And spent the proceeds on a telescope
To satisfy a lifelong curiosity
About our place among the infinities . . .

The Poetry of Robert Frost
The Star Splitter
l. 16–19

The telescope at one end of his beat,
And at the other end the microscope.

The Poetry of Robert Frost
The Bear
l. 18–19

Galilei, Galileo
O telescope, instrument of much knowledge, more precious than any scepter! Is not he who holds thee in his hand made king and lord of the works of God?

Quoted by William H. Jefferys and R. Robert Robbins in
Discovering Astronomy (p. 174)

Gamow, George
My telescope
Has dashed your hope;
   Your tenets are refuted.
Let me be terse:
Our universe
   Grows daily more diluted!

Mr. Tompkins in Paperback
Chapter 6 (p. 63)

Herschel, Sir John
In the old Telescope's tube we sit,
And the shades of the past around us flit . . .

Herschel at the Cape (p. xix)

Hubble, Edwin
With increasing distance, our knowledge fades, and fades rapidly. Eventually we reach the dim boundary—the utmost limits of our

telescopes. There, we measure shadows, and we search among ghostly errors of measurement for landmarks that are scarcely more substantial.

The Realm of the Nebulae (p. 202)

Kitchiner, William
Immense telescopes are only about as useful as the enormous spectacles which are suspended over the doors of opticians!

Quoted by William Sheehan in
Planets & Perception (p. 113)

Longair, Malcolm
'Twas brillig and the slithy toves
Brought plans of telescopes fair to see.
The Jabberwock, he clapped his hands
And said, "That's just for me."

Alice and the Space Telescope
Chapter 2 (p. 7)

Milton, John
. . . a spot like which perhaps
Astronomer in the Sun's lucent Orbe
Through his glaz'd Optic Tube yet never saw.

Paradise Lost
Book III, l. 588–90

Newton, Sir Isaac
If the Theory of making Telescopes could at length be fully brought into Practice, yet there would be a certain Bound beyond which Telescopes could not perform.

Opticks
Book 1, Part 1, Proposition viii, problem 2

Noyes, Alfred
"To-morrow night"—so wrote the chief—
  "we try
Our great new telescope, the hundred-inch.
Your Milton's 'optic tube' has grown in
  power . . . "

Watchers of the Sky
Prologue (p. 2)

My periwig's askew, my ruffle stained
With grease from my new telescope!

Watchers of the Sky
William Herschel Conducts (p. 231)

TEMPERATURE

Smith, H. Allen

The radiation falling on heaven will heat it to the point where the heat lost by radiation is just equal to the heat received by radiation. In other words, heaven loses fifty times as much heat as the earth by radiation. Using the well-known Stefan–Boltzmann fourth-power law for radiation $(H/E)^4 = 50$ where H is the absolute temperature of heaven and E is the absolute temperature of the earth—300 °C (273 plus 27). This gives H as 798° absolute or 525 °C.

The exact temperature cannot be computed, but it must be less than 444.6 °C, the temperature at which brimstone or sulphur changes from a liquid to a gas . . .. If it were above this point it would be a vapor and not a lake.

We have then the following: Temperature of heaven, 525 °C or 977 °F. Temperature of hell, less than 445 °C or less than 833 °F. Therefore, heaven is hotter than hell.

Saturday Review
From Martin Levin's
Phoenix Nest
May 21, 1960

Unknown

The temperature of Heaven can be rather accurately computed. Our authority is Isaiah 30:26, "Moreover, the light of the Moon shall be as the light of the Sun and the light of the Sun shall be sevenfold, as the light of seven days." Thus Heaven receives from the Moon as much radiation as we do from the Sun, and in addition 7 × 7 (49) times as much as the Earth does from the Sun, or 50 times in all. The light we receive from the Moon is one 1/10,000 of the light we receive from the Sun, so we can ignore that . . .. The radiation falling on Heaven will heat it to the point where the heat lost by radiation is just equal to the heat received by radiation, i.e., Heaven loses 50 times as much heat as the Earth by radiation. Using the Stefan–Boltzmann law for radiation, $(H/E)^4 = 50$, where E is the absolute temperature of the earth (300 K), gives H as 798 K (525 °C). The exact temperature of Hell cannot be computed . . .. [However] Revelations 21:8 says "But the fearful, and unbelieving . . . shall have their part in the lake which burneth with fire and brimstone." A lake of molten brimstone means that its temperature must be at or below the boiling point, 444.6 °C. We have, then, that Heaven, at 525 °C is hotter than Hell at 445 °C.

Applied Optics
HEAVEN IS HOTTER THAN HELL (p. A14)
Volume 11, Number 8, August 1972

There is an appointed time for everything and a time for every affair under the heavens.

Ecclesiasties – (See p. 321)

THEORY

Adams, Douglas
There is a theory which states that if ever anyone discovers exactly what the Universe is for and why it is here, it will instantly disappear and be replaced by something even more bizarre and inexplicable.

There is another which states that this has already happened.

The Restaurant at the End of the Universe (Beginning of book)

Berkeley, Edmund C.
The World is more complicated than most of our theories make it out to be.

Computers and Automation
Right Answers—A Short Guide for Obtaining Them
September 1969 (p. 20)

Bernard, Claude
A theory is merely a scientific idea controlled by experiment.

An Introduction to the Study of Experimental Medicine (p. 26)

Bohr, Niels
. . . your theory is crazy. The question which divides us is whether it is crazy enough to have a chance of being correct.

Quoted by Martin Gardner in
The Ambidextrous Universe (p. 280)

Your theory is crazy—but not crazy enough to be true.

Quoted by Arthur C. Clarke in
The Lost Worlds of 2001
Chapter 30

Born, Max
A theory, to be of any real use to us, must satisfy two tests. In the first place, it must not make use of any ideas which are not confirmed by experiment. Special assumptions must not be dragged in merely to meet some particular difficulty. In the second place, the theory must not only explain all the facts known already, but must also enable us to foresee other facts which were not known before and can be tested by further experiment.

The Restless Universe (pp. 5–6)

Bridgman, P.W.
Every new theory as it arises believes in the flush of youth that it has the long-sought goal; it sees no limits to its applicability, and believes that at long last it is the fortunate to achieve the "right" answer.

The Nature of Physical Theory (p. 136)

Buckland, Frank
Your theory is most excellent, and I shall endeavor to collect facts for you with a view to its elucidation.

Quoted by Karl Pearson in
The Life, Letters, and Language of Francis Galton
Volume II (p. 87)

Chan, Charlie
Theory like mist on eyeglasses. Obscure facts.

In the movie *Charlie Chan in Egypt*

Chan, Jimmy
But Pop, I've got a theory!

In the movie *Charlie Chan in Panama*

Chesterton, Gilbert Keith
A man warmly concerned with any large theories has always a relish for applying them to any triviality.

The Wisdom of Father Brown
The Absence of Mr. Glass

Clarke, Arthur C.
"I'd be glad to settle without the theory," remarked Kimball, "if I could even understand what this thing is—or what it's supposed to do."

The Lost Worlds of 2001
Chapter 30

Colton, Charles Caleb
Theory is worth little, unless it can explain its own phenomena, and it must effect this without contradicting itself; therefore, the facts are sometimes assimilated to the theory, rather than the theory to the facts.

Lacon (p. 77)

Professors in every branch of the sciences prefer their own theories to truth: the reason is, that their theories are private property, but truth is common stock.

Lacon (p. 189)

Couderc, Paul
The only conceptions that succumb are those that pretend to fix the image of a profound reality: true relations among things survive, united to the true new relations in the burgeoning theory. Let us then rejoice at the massacre of old theories because this is the criterion of progress. There is, I think, no ground for fear that nature will undernourish the seekers. Nothing should diminish our enthusiasm for the experimental victories, decisive and definitive, of the past thirty years.

Quoted by Lucienne Felix in
The Modern Aspect of Mathematics (p. 31)

Da Vinci, Leonardo
The supreme misfortune is when theory outstrips performance.

Notebooks
c

Darwin, Charles
. . . for without the making of theories I am convinced there would be no observation.

In Francis Darwin's
The Life and Letters of Charles Darwin
Volume II
C. Darwin to C. Lyell
June 1st [1860] (p. 108)

Davies, J.T.
. . . a theory arises from a leap of the imagination . . .

The Scientific Approach (p. 11)

Theories are generalizations and unifications, and as such they cannot logically follow only from our experiences of a few particular events. Indeed we often generalize from a single event, just as a dog does who, having once seen a cat in a certain driveway, looks eagerly around whenever he passes that place in future. Evidently this latter activity is equivalent to testing the theory . . . that "there is always a cat in that driveway".

The Scientific Approach (p. 11)

Duhem, Pierre

A physical theory is not an explanation. It is a system of mathematical propositions, deduced from a small number of principles, which aim to represent as simply, as completely, and as exactly as possible a set of experimental laws.

The Aim and Structure of Physical Theory (p. 19)

Eddington, Sir Arthur Stanley

The relativity theory of physics reduces everything to relations; that is to say, it is structure, not material, which counts. The structure cannot be built up without material; but the nature of the material is of no importance.

Space, Time, and Gravitation (p. 197)

We have found a strange footprint on the shores of the unknown. We have devised profound theories, one after another to account for its origin. At last, we have succeeded in reconstructing the creature that made the footprint. And lo! It is our own.

Space, Time and Gravitation (p. 201)

. . . a reasoned theory is preferable to blind extrapolation.

The Expanding Universe (p. 18)

There was just one place where the theory did not seem to work properly, and that was—infinity. I think Einstein showed his greatness in the simple and drastic way in which he disposed of difficulties at infinity. He abolished infinity. He slightly altered his equations so as to make space at great distances bend round until it closed up. So that, if in Einstein's space you kept going right on in one direction, you do not get to infinity; you find yourself back at your starting-point again. Since there was no longer any infinity, there could be no difficulties at infinity. Q.E.D.

The Expanding Universe (pp. 21–2)

Einstein, Albert

A theory is the more impressive the greater the simplicity of its premises, the more different kinds of things it relates, and the more extended its area of applicability.

Albert Einstein: Autobiographical Notes (p. 31)

I have learned something else from the theory of gravitation: no collection of empirical facts however comprehensive can ever lead to the setting up of such complicated equations. A theory can be tested by experience, but there is no way from experience to the construction of a theory. Equations of such complexity as are the equations of the gravitational field can be found only through the discovery of a logically simple mathematical condition that determines the equations completely or almost completely. Once one has obtained those sufficiently strong formal conditions, one requires only little knowledge of facts for the construction of the theory; in the case of the equations of gravitation it is the four-dimensionality and the symmetric tensor as expression for the structure of space that, together with the invariance with respect to the continuous transformation group, determine the equations all but completely.

Albert Einstein: Autobiographical Notes (p. 85)

If my theory of relativity is proven successful, Germany will claim me as a German and France will declare that I am a citizen of the world. Should my theory prove untrue, France will say that I am a German, and Germany will declare that I am a Jew.

The Great Quotations (p. 226)
Address at the Sorbonne

From a systematic theoretical point of view, we may imagine the process of evolution of an empirical science to be a continuous process of induction. Theories are evolved and are expressed in short compass as statements of a large number of individual observations in the form of empirical laws, from which the general laws can be ascertained by comparison. Regarded in this way, the development of a science bears some resemblance to the compilation of a classified catalogue. It is, as it were, a purely empirical enterprise.

Relativity
Appendix III
The Experimental Confirmation of the General Theory of Relativity

It is the theory that decides what we can observe.

Quoted by Werner Heisenberg in
Physics and Beyond: Encounters and Conversations (p. 63)

Creating a new theory is not like destroying an old barn and erecting a skyscraper in its place. It is rather like climbing a mountain, gaining new and wider views, discovering unexpected connections between our starting point and its rich environment. But the point from which we started out still exists and can be seen, although it appears smaller and forms a tiny part of our broad view gained by the mastery of the obstacles on our adventurous way up.

The Evolution of Physics (pp. 158–9)

Eliot, George
The possession of an original theory which has not yet been assailed must certainly sweeten the temper of a man who is not beforehand ill-natured.

Theophrastus Such
How We Encourage Research (p. 26)

Fermat, Pierre
We have found a beautiful and most general proposition, namely, that every integer is either a square, or the sum of two, three or at most four squares. This theorem depends on some of the most recondite mysteries of number, and is not possible to present its proof on the margin of this page.

Quoted by Tobias Dantzig in
Number: The Language of Science (p. 269)

I have found a very great number of exceedingly beautiful theorems.

Quoted by E.T. Bell in
Men of Mathematics (p. 56)

Gilbert, W.S.
Sullivan, Arthur
About binomial theorem I'm teeming with a lot o' news—
With many cheerful facts about the square of the hypotenuse.

The Pirates of Penzance
Act I

Goethe, Johann Wolfgang von
Dear friend, all theory is grey
And green the golden tree of life.

Faust
The First Part
Study, l. 2038–9

Hamilton, Edith
Theories that go counter to the facts of human nature are foredoomed.

The Roman Way
Comedy's Mirror

Heisenberg, Werner
Modern theory did not arise from revolutionary ideas which have been, so to speak, introduced into the exact sciences from without. On the contrary, they have forced their way into research which was attempting

consistently to carry out the program of classical physics they arise out of its very nature.

Quoted by Heinz R. Pagels in
The Cosmic Code (p. 67)

Hertz, H.
To this question, "What is Maxwell's theory?" I cannot give any clearer or briefer answer than the following: Maxwell's theory is the system of Maxwell's equations.

Electric Waves: Being Researches on the Propagation of Electric
Action with Finite Velocity through Space (p. 23)

Hilton, James
And I believe that the Binomial Theorem and a Bach Fugue are, in the long run, more important than all the battles of history.

This Week Magazine
1937

Holmes, Sherlock
"One forms provisional theories and waits for time or fuller knowledge to explode them. A bad habit, Mr. Ferguson, but human nature is weak."

In Arthur Conan Doyle's
The Complete Sherlock Holmes
The Adventure of the Sussex Vampire

I don't mean to deny that the evidence is in some ways very strong in favour of your theory, I only wish to point out that there are other theories possible.

In Arthur Conan Doyle's
The Complete Sherlock Holmes
Adventures of the Norwood Builder

Hubble, Edwin
No theory is sacred.

The Nature of Science and other Lectures
Experiment and Experience (p. 41)

Kitaigorodski, Aleksander Isaakovich
A first-rate theory predicts; a second-rate theory forbids and a third-rate theory explains after the event.

Lecture, ICU Amsterdam, August 1975

Lichtenberg, Georg Christoph
This whole theory is good for nothing except disputing about.

Lichtenberg: Aphorisms & Letters
Aphorisms (p. 57)

Mach, Ernst

The object of natural science is the connection of phenomena; but the theories are like dry leaves which fall away when they have long ceased to be the lungs of the tree of science.

History and Root of the Principle of the Conservation of Energy (p. 74)

Maimonides, Moses

Man knows only these poor mathematical theories about the heavens, and only God knows the real motions of the heavens and their causes.

Quoted by Phillip Frank in
Modern Science and its Philosophy (p. 222)

Mayes, Harlan Jr.

Theory, glamorous mother of the drudge experiment.

Quoted by Eric M. Rogers in
Physics for the Inquiring Mind
Chapter 40 (p. 648)

Nietzsche, Friedrich

It is certainly not the least charm of a theory that it is refutable . . .

Beyond Good and Evil
Chapter I, Section 18

Nizer, Louis

The argument seemed sound enough, but when a theory collides with a fact, the result is a tragedy.

My Life in Court (p. 433)

Novalis

Theories are like fishing: it is only by casting into unknown waters that you may catch something.

Quoted by Jean-Pierre Luminet in
Black Holes (p. 1)

Oman, John

To refuse to consider any possibility is merely the old habit of making theory the measure of reality.

The Natural and the Supernatural
Chapter XV (p. 269)

Petit, Jean-Pierre

Sir, please believe me, it's the first time this has ever happened. Have another try, don't get upset. You know our Theorems are GUARANTEED.

Euclid Rules OK? (p. 11)

Poincaré, Henri

At the first blush it seems to us that theories last only a day and that ruins upon ruins accumulate . . ... But if we look more closely, we see that what thus succumb are the theories properly so called, those which pretend to teach us what things are. But there is in them something which usually survives. If one of them taught us a true relation, this relation is definitively acquired, and it will be found again under a new disguise in the other theories which will successively come to reign in place of the old.

The Foundations of Science
The Value of Science
Science and Reality (p. 351)

Popper, Karl R.

But this is sufficient to show that a high probability cannot be one of the aims of science. For the scientist is most interested in theories with a high content. He does not care for highly probable trivialities but for bold and severely testable (and severely tested) hypotheses. If (as Carnap tells us) a high degree of confirmation is one of the things we aim at in science, then degree of confirmation cannot be identified with probability.

This may sound paradoxical to some people. But if high probability were an aim of science, then scientists should say as little as possible, and preferably utter tautologies only. But their aim is to "advance" science, that is to add to its content. Yet this means lowering its probability. And in view of the high content of universal laws, it is neither surprising to find that their probability is zero . . .

Conjectures and Refutations: The Growth of Scientific Knowledge (p. 286)

Theories are nets cast to catch what we call "the world": to rationalize, to explain, and to master it. We endeavor to make the mesh ever finer and finer.

The Logic of Scientific Discovery (p. 59)

Rainich, G.Y.

Moreover, the really fundamental things have a way of appearing to be simple once they have been stated by a genius, who was in this case Minkowski..

Bulletin of the American Mathematical Society
Analytic Function and Mathematical Physics (p. 700)
October 1931

Reichenbach, Hans

The study of inductive inference belongs to the theory of probability, since observational facts can make a theory only probable but will never make it absolutely certain.

The Rise of Scientific Philosophy (p. 231)

Robinson, Arthur
In short, quantum mechanics, special relativity, and realism cannot all be true.

Science
Quantum Mechanics Passes Another Test (p. 435)
Volume 217, Number 4558, July 30, 1982

Romanoff, Alexis L.
A theory is worthless without good supporting data.

Encyclopedia of Thoughts
Aphorisms
2410

Sayers, Dorothy L.
Very dangerous things, theories.

The Unpleasantness at the Bellona Club
Chapter 4

Seeger, Raymond J.
It is noteworthy that the etymological root of the word theatre is the same as that of the word theory, namely a view. A theory offers us a better view.

Journal of the Washington Academy of Sciences
Volume 36, 1946 (p. 286)

Thomson, William [Lord Kelvin]
Fourier's Theorem is not only one of the most beautiful results of modern analysis, but it may be said to furnish an indispensable instrument in the treatment of nearly every recondite question in modern physics.

Quoted by E.T. Bell in
Men of Mathematics (p. 183)

Unknown
If you have to prove a theorem, do not rush. First of all, understand fully what the theorem says, try to see clearly what it means. Then check the theorem; it could be false. Examine the consequences, verify as many particular instances as are needed to convince yourself of its truth. When you have satisfied yourself that the theorem is true, you start proving it.

Source unknown

Schumpter's Observation of Scientific Theories. Any theory can be made to fit any facts by means of appropriate additional assumptions.

Quoted by Paul Dickson in
The Official Rules (S-165)

The Treadmill Theorem states that every solution entails k GE 1 $[k \geqslant 1]$ new problems.

Source unknown

The Dirty Data Theorem states that "real world" data tend to come from bizarre and unspecifiable distributions of highly correlated variables and have unequal sample sizes, missing data points, non-independent observations, and an indeterminate number of inaccurately recorded values.

Source unknown

Why bother to make it elegant if it already works?

Source unknown

Capturing Lions

The following represent several mathematical methods for capturing a lion in the middle of the Sahara Desert.

The Dirac method
We observe that wild lions are, *ipso facto*, not observable in the Sahara Desert. Consequently, if there are any lions in the Sahara, they are tame. The capture of a tame lion may be left as an exercise for the reader.

The Schrödinger method
At any given moment there is a positive probability that there is a lion in the cage. Sit down and wait.

Place a tame lion in the cage, and apply a Majorana exchange operator between it and a wild lion.

As a variant, let us suppose, to fix ideas, that we require a male lion. We place a tame lioness in the cage, and apply a Heisenberg exchange operator which exchanges the spins.

A relativistic method
We distribute about the desert lion bait containing large portions of the Companion of Sirius. When enough bait has been taken, we project a beam of light across the desert. This will bend right around the lion, who will then become so dizzy that he can be approached with impunity.

The thermodynamical method
We construct a semi-permeable membrane, permeable to everything except lions, and sweep across the desert.

We irradiate the desert with slow neutrons. The lion becomes radioactive, and a process of disintegration sets in. When the decay has proceeded sufficiently far, he will become incapable of showing fight.

The magneto-optical method
We plant a large lenticular bed of catnip (*Nepeta catarial*) whose axis lies along the direction of the horizontal component of the earth's magnetic field, and place a cage at one of its foci. We distribute over the desert large quantities of magnetized spinach (*Spinacia oleraceal*) which, as is well known, has a high ferric content. The spinach is eaten by the herbivorous denizens of the desert, which are in turn eaten by lions. The lions are then oriented parallel to the earth's magnetic field, and the resulting beam of lions is focused by catnip upon the cage.

Quoted by H. Pétard in
American Mathematical Monthly
A Contribution to the Mathematcial Theory of Big Game Hunting
Volume 45, Number 7 (p. 446) 1938

Facts without theory is trivia. Theory without facts is bullshit.

Source unknown

Voltaire
"Let us work without theorising," said Martin; "tis the only way to make life endurable."

Candide
XXX

Weinberg, Steven
This is often the way it is in physics—our mistake is not that we take our theories too seriously, but that we do not take them seriously enough.

The First Three Minutes (p. 131)

Whitehead, Alfred North
On the absolute theory, bare space and bare time are such very odd existences, half something and half nothing.

Proceedings of the Aristotelian Society
The Idealistic Interpretations of Einstein's Theory (p. 131)
N.S. Vol. XXII, Part III

Wolfowitz, J.
Except perhaps for a few of the deepest theorems, and perhaps not even these, most of the theorems of statistics would not survive in mathematics if the subject of statistics itself were to die out. In order to survive the subject must be more responsive to the needs of application.

Essays in Probability and Statistics (p. 748)

THERMODYNAMICS

Bohr, Niels
The old thermodynamics . . . is to statistical thermodynamics what classical mechanics is to quantum mechanics.

Quoted by Werner Heisenberg in
Physics and Beyond: Encounters and Conversations (p. 107)

Cardenal, Ernesto
The second law of thermodynamics!:
energy is indestructible in quantity
but continually changes in form.
    And it always runs down like water.

Cosmic Canticle
Cantiga 3
Autumn Fugue

Eddington, Sir Arthur Stanley
The law that entropy always increases—the Second Law of Thermodynamics—holds, I think, the supreme position among the laws of Nature. If someone points out to you that your pet theory of the universe is in disagreement with Maxwell's equations—then so much the worse for Maxwell's equations. If it is found to be contradicted by observation— well, these experimentalists do bungle things sometimes. But if your theory is found to be against the Second Law of Thermodynamics I can give you no hope; there is nothing for it but to collapse in deepest humiliation.

The Nature of the Physical World
Chapter IV
Coincidences (p. 74)

Meixner, J.
A careful study of the thermodynamics of electrical networks has given considerable insight into these problems and also produced a very interesting result: the non-existence of a unique entropy value in a state

318

which is obtained during an irreversible process . . .. I would say, I have done away with entropy. The next step might be to let us also do away with temperature.

In Edward B. Stuart, Benjamin Gal-Or and Alan J. Brainard (Editors)
A Critical Review of Thermodynamics

Stenger, Victor J.
Scientists speak of the Law of Inertia or the Second Law of Thermodynamics as if some great legislature in the sky once met and set down rules to govern the universe.

Not By Design (p. 14)

We hope to explain the entire universe in a single, simple formula that you can wear on your T-shirt.

Leon Lederman – (See p. 341)

TIME

Barrow, Isaac
Because Mathematicians frequently make use of Time, they ought to have a distinct idea of the meaning of that Word, otherwise they are Quacks . . .

Quoted by Paul Davies in
About Time (p. 183)

Bergson, Henri
Time is invention or it is nothing at all.

Creative Evolution (p. 341)

Bohm, David
Eternity can be affected by what happens in time.

Quoted by Renée Weber in
Dialogues with Scientists and Sages (p. 91)

But the puzzle is, what happened before time began?

Quoted by Renée Weber in
Dialogues with Scientists and Sages (p. 199)

Bondi, Sir Hermann
Time must never be thought of as pre-existing in any sense; it is a manufactured quantity.

Quoted by Paul Davies in
About Time (p. 22)

Borges, Jorge Luis
Differing from Newton and Schopenhauer . . . Ts'ui Pen did not think of time as absolute or uniform. He believed in an infinite series of times, in a dizzily growing, ever spreading network of diverging, converging and parallel times. This web of time—the strands of which approach one another, bifurcate, intersect or ignore each other through the centuries—embraces every possibility.

Cuentos De Jorge Luis Borges
El Jardín De Senderos Que Se Bifurcan (p. 43)

Time is a river which sweeps me along, but I am the river; it is a tiger which mangles me, but I am the tiger; it is a fire which consumes me, but I am the fire. The world, unfortunately, is real; I, unfortunately, am Borges.

A Personal Anthology
A New Refutation of Time (p. 64)

Carroll, Lewis

Alice sighed wearily. "I think you might do something better with the time," she said, "than wasting it in asking riddles with no answers."

"If you knew Time as well as I do," said the Hatter, "you wouldn't talk about wasting *it*."

The Complete Works of Lewis Carroll
Alice's Adventures in Wonderland
The Mad Tea Party

"Ah! that accounts for it," said the Hatter. "He won't stand beating. Now, if you only kept on good terms with him, he'd do almost anything you liked with the clock . . . but you could keep it to half-past one as long as you liked."

The Complete Works of Lewis Carroll
Alice's Adventures in Wonderland
The Mad Tea Party

Clemence, G.M.

The measurement of time is essentially a process of counting.

The American Scientist
Time and its Measurement (p. 261)
Volume 40, Number 2, April 1952

Cleugh, Mary F.

The ghost of time cannot permanently be laid.

Time and Its Importance in Modern Thought (p. 16)

Ecclesiasties 3:1

There is an appointed time for everything and a time for every affair under the heavens.

The Bible

Eddington, Sir Arthur Stanley

In any attempt to bridge the domains of experience belonging to the spiritual and physical sides of our nature, time occupies the key position.

The Nature of the Physical World
Chapter V
Linkage of Entropy with Becoming (p. 91)

The great thing about time is that it goes on.

Quoted by Paul Davies in
About Time (p. 25)

Einstein, Albert
Michele has left this strange world just before me. This is of no importance. For us convinced physicists the distinction between past, present and future is an illusion, although a persistent one.

Quoted by Eric J. Lerner in
The Big Bang Never Happened (p. 283)

The distinction between past, present and future is only an illusion, even if a stubborn one.

Quoted by Paul Davies in
About Time (p. 70)

Eliot, T.S.
Time present and time past
Are both perhaps present in time future,
And time future contained in time past.
If all time is eternally present
All time is unredeemable.

Burnt Norton
I

Ford, John
'A has shook hands with time.

The Broken Heart
Act V, scene ii, l. 156

Fraser, J.T.
The resulting dichotomy between time felt and time understood is a hallmark of scientific–industrial civilization, a sort of collective schizophrenia.

Quoted by Eric J. Lerner in
The Big Bang Never Happened (p. 283)

Gardner, Earl Stanley
Time is really nothing but a huge circle. You divide a circle of three hundred and sixty degrees into twenty-four hours, and you get fifteen degrees of arc that is the equivalent of each hour.

The Case of the Buried Clock (p. 82)

Heinlein, Robert A.
The hardest part about gaining any new idea is sweeping out the false idea occupying that niche. As long as that niche is occupied, evidence

and proof and logical demonstration get nowhere. But once the niche is emptied of the wrong idea that has been filling it— once you can honestly say, "I don't know," then it becomes possible to get at the truth.

The Cat Who Walks Through Walls (p. 244)

Heracleitus
Time is like a river flowing endlessly through the universe.

Quoted by Franzo H. Crawford in
Introduction to the Science of Physics (p. 160)

Hoyle, Fred
If there is one thing we can be sure enough of in physics it is that all time exists with equal reality.

October the First Is Too Late (p. 75)

Jeans, Sir James Hopwood
. . . time figures as the mortar which binds the bricks of matter together . . .

The Mysterious Universe
Into the Deep Waters (p. 121)

Kant, Immanuel
Time is not an empirical conception. For neither coexistence nor succession would be perceived by us, if the representation of time did not exist as a foundation *a priori*. Without this presupposition we could not represent to ourselves that things exist together at one and the same time, or at different times, that is, contemporaneously, or in succession.

The Critique of Pure Reason
First Part
Of Time
Metaphysical Exposition of this Conception, 5

Milton, John
Fly envious *Time*, till thou run out thy race . . .

The Complete Poetical Works of John Milton
On Time
l. 1

Misner, Charles W.
Thorne, Kip S.
Wheeler, John A.
Time is defined so that motion looks simple.

Gravitation
Time (p. 23)

Nabokov, Vladimir
Pure Time, Perceptual Time, Tangible Time, Time free of content, context, and running commentary—this is my time and theme. All the rest is numerical symbol or some aspect of Space. The texture of Space is not that of Time, and the piebald four-dimensional sport bred by relativists is a quadruped with one leg replaced by the ghost of a leg. My time is also Motionless Time (we shall presently dispose of "flowing" time, water-clock time, water-closet time).

Ada or Ardor: A Family Chronicle (p. 574)

Newton, Sir Isaac
I do not define time, space, place and motion, as being well known to all.

Mathematical Principles of Natural Philosophy
Definitions, Definition VIII, Scholium

Absolute, true, and mathematical time, of itself, and from its own nature, flows equably without relation to anything external.

Mathematical Principles of Natural Philosophy
Definitions, Definition VIII, Scholium I

Ovid
Time glides by with constant movement, not
  unlike a stream.
For neither can a stream stay its course, nor can
  the fleeting hour.

Metamorphoses
XV, 180

Penrose, Roger
Our present picture of physical reality, particularly in relation to the nature of time, is due for a grand shake-up—even greater, perhaps, than that which has already been provided by present-day relativity and quantum mechanics.

The Emperor's New Mind (p. 371)

Plato
Time, then, and the heaven came into being at the same instant in order that, having been created together, if ever there was to be a dissolution of them, they might be dissolved together.

Timaeus
Section 38

Prigogine, Ilya
The irreversibility [of time] is the mechanism that brings order out of chaos.

Quoted by Eric J. Lerner in
The Big Bang Never Happened (p. 283)

Time is creation. The future is just not there.

Quoted by Eric J. Lerner in
The Big Bang Never Happened (p. 321)

The Rocky Horror Picture Show
With a bit of a mind flip
You're into the time slip
And nothing can ever be the same.

The Time Warp

Saint Augustine
Time is like a river made up of events which happen, and its current is strong; no sooner does anything appear than it is swept away.

Quoted by Paul Davies in
Other Worlds (p. 186)

How can the past and future be when the past no longer is and the future is not yet? As for the present, if it were always present and never moved on to become the past, it would not be time but eternity.

Confessions (p. 253)

Stevenson, R.L.
"She is settling fast," said the First Lieutenant as he returned from shaving.

"*Fast*, Mr. Spoker?" asked the Captain. "The expression is a strange one, for Time (if you will think of it) is only relative."

Fables
The Sinking Ship

Thoreau, Henry David
Time is but the stream I go a-fishin in.

Walden
Where I Lived, and What I Lived For

Unknown
Time is just one damn thing after another.

Source unknown

Time is God's way of keeping things from happening all at once.

Source unknown

When a snail crossed the road, he was run over by a turtle. Regaining consciousness in the emergency room, he was asked what caused the accident. "I really can't remember," the snail replied. "You see, it all happened so fast."

Source unknown

Virgil
Time is flying—flying never to return.

Quoted by Paul Davies in
Other Worlds (p. 186)

Vyasa
Time does not sleep when all things sleep,
Only Time stands straight when all things fall.
Is, was, and shall be are Time's Children.
Is, was, and shall be are Time's Children.

The Mahabharata of Vyasa
The Beginning (p. 65)

Wheeler, John A.
Of all the obstacles to a thoroughly penetrating account of existence, none looms up more dismayingly than "time." Explain time? Not without explaining existence. Explain existence? Not without explaining time. To uncover the deep and hidden connection between time and existence, to close on itself our quartet of questions, is a task for the future.

Quoted by Eugene F. Mallove in
The Quickening Universe (p. 189)

Should we be prepared to see some day a new structure for the foundations of physics that does away with time?

Quoted by Paul Davies in
About Time (p. 178)

Time ends. That is the lesson of the "big bang." It is also the lesson of the black hole.

Proceedings of the American Philosophical Society
The Lesson of the Black Hole (p. 25)
Volume 125

Time is nature's way to keep everything from happening at once.

Quoted by Paul Davies in
About Time (p. 236)

Whitehead, Alfred North

It is impossible to meditate on time and the mystery of the creative passage of nature without an overwhelming emotion at the limitations of human intelligence.

The Concept of Nature (p. 73)

Wilford, John Noble

Scientists report profound insight on how time began

New York Times
Friday April 24, 1992
Section A, page 1

Yeats, William Butler

Time drops in decay,
Like a candle burnt out.

The Collected Poems of W.B. Yeats
The Moods

HMM, MY WATCH APPEARS TO HAVE STOPPED....

. . . time figures as the mortar which binds the bricks of matter together . . .

Sir James Hopwood Jeans – (See p. 323)

TRUTH

Aurobindo, Ghose

. . . the intellect is incapable of knowing the supreme Truth; it can only range about seeking Truth, and catching fragmentary representations of it, not the thing itself, and trying to piece them together.

The Riddle of the World (p. 23)

Bacon, Francis

Truth emerges more readily from error than from confusion.

Quoted by Ritchie Calder in
Man and the Cosmos (p. 19)

Bohr, Niels

The opposite of a correct statement is a false statement. But the opposite of a profound truth may well be another profound truth.

Quoted by Werner Heisenberg in
Physics and Beyond (p. 102)

Heaviside, Oliver

We do not dwell in the Palace of Truth. But, as was mentioned to me not long since, "There is a time coming when all things shall be found out." I am not so sanguine myself, believing that the well in which Truth is said to reside is really a bottomless pit.

Electromagnetic Theory
Chapter I
Volume I (p. 1)

Lawson, Alfred William

Education is the science of knowing TRUTH.
Miseducation is the art of absorbing FALSITY.
TRUTH is that which is, not that which ain't.
FALSITY is that which ain't, not that which is.

Quoted by Martin Gardner in
Fads and Fallacies (p. 76)

Levy, Hyman

Truth is a dangerous word to incorporate within the vocabulary of science. It drags with it, in its train, ideas of permanence and immutability that are foreign to the spirit of a study that is essentially an historically changing movement, and that relies so much on practical examination within restricted circumstances . . .. Truth is an absolute notion that science, which is not concerned with any such permanency, had better leave alone.

The Universe of Science (pp. 206–7)

Newton, Sir Isaac

I do not know what I may appear to the world, but to myself I seem to have been only like a boy playing on the sea-shore, and diverting myself in now and then finding a smoother pebble or a prettier shell than ordinary, whilst the great ocean of truth lay all undiscovered before me.

Quoted by David Brewster in
Memoirs of the Life, Writings and Discoveries of Sir Isaac Newton
Chapter 27 (p. 331)

Planck, Max

A new scientific truth does not triumph by convincing its opponents and making them see the light, but rather because its opponents eventually die, and a new generation grows up that is familiar with it.

Scientific Autobiography (pp. 33–4)

Reichenbach, Hans

He who searches for truth must not appease his urge by giving himself up to the narcotic of belief.

Quoted by Ruth Renya in
The Philosophy of Matter in the Atomic Era (p. 16)

Renan, Ernest

The simplest schoolboy is now familiar with truths for which Archimedes would have sacrificed his life.

Quoted by L.I. Ponomarev in
The Quantum Dice (p. 34)

Spencer-Brown, George

To arrive at the simplest truth, as Newton knew and practiced, requires years of *contemplation*. Not activity. Not reasoning. Not calculating. Not busy behavior of any kind. Not reading. Not talking. Not making an effort. Not thinking. Simply *bearing in mind* what it is one needs to know. And yet those with the courage to tread this path to real discovery are not only offered practically no guidance on how to do so, they are actively discouraged and have to set about it in secret, pretending meanwhile to be diligently engaged in the frantic diversions and to conform with the deadening personal opinions which are continually being thrust upon them.

Laws of Form
Appendix I (p. 110)

Wilde, Oscar

It is a terrible thing for a man to find out suddenly that all his life he has been speaking nothing but the truth.

Quoted by John D. Barrow in
The World within the World (p. 260)

JACK — . . . That, my dear Algy, is the whole truth, pure and simple.

ALGERNON — The truth is rarely pure and never simple.

The Importance of Being Earnest
Act I

Wilkins, John

That the strangeness of this opinion is no sufficient reason why it should be rejected, because other certain truths have been formerly esteemed ridiculous, and great absurdities entertayned by common consent.

The Discovery of a World in the Moone (p. 1)

UNCERTAINTY

Emilsson, Tryggvi

Historians have concluded that W. Heisenberg must have been contemplating his love life when he discovered the Uncertainty Principle: —When he had the time, he didn't have the energy and, —When the moment was right, he couldn't figure out the position.

Quoted on the Internet

Heisenberg, Werner

Even when this arbitrariness is taken into account the concept "observation" belongs, strictly speaking, to the class of ideas borrowed from the experiences of everyday life. It can only be carried over to atomic phenomena when due regard is paid to the limitations placed on all space–time descriptions by the uncertainty principle.

The Physical Principles of the Quantum Theory (p. 64)

In fact, our ordinary description of nature, and the idea of exact laws, rests on the assumption that it is possible to observe the phenomena without appreciably influencing them.

The Physical Principles of the Quantum Theory (p. 62)

Unknown

There once was a man who said: "Damn!
I can't possibly be in this tram
For how can I know
Both how fast that I go
And also the place where I am."

Source unknown

Heisenberg might have slept here.

Source unknown

UNIVERSE

Alfvén, Hannes
I have never thought that you can get the extremely clumpy, heterogeneous universe we have today from a smooth and homogenous one dominated by gravitation.

Quoted by Eric J. Lerner in
The Big Bang Never Happened (p. 42)

Allen, Woody
Can we actually "know" the universe? My God, it's hard enough finding your way around in Chinatown.

Getting Even (p. 29)

The universe is merely a fleeting idea in God's mind—a pretty uncomfortable thought, particularly if you've just made a down payment on a house.

Getting Even (p. 33)

Archimedes
There are some, King Gelon, who think that the number of the sand is infinite in multitude; and I mean by the sand not only that which exists about Syracuse and the rest of Sicily but also that which is found in every region inhabited or uninhabited. Again there are some who, without regarding it as infinite, yet think that no number has been named which is great enough to exceed its multitude . . . .. But I will try to show you by means of geometrical proofs, which you will be able to follow, that, of the numbers named by me, and given in the work which I sent to Zeuxippus, some exceed not only the number of the mass of sand equal in magnitude to the earth, but also that of a mass equal in magnitude to the universe.

The Works of Archimedes
The Sand-Reckoner

Bacon, Leonard
Eddington's universe goes phut.
Richard Tolman's can open and shut.
Eddington's bursts without grace or tact,
But Tolman's swells and perhaps may contract

Rhyme and Punishment
Richard Tolman's Universe

Bagehot, Walter
Taken as a whole, the universe is absurd. There seems an unalterable contradiction between the human mind and its employments.

Literary Studies
Volume I (p. 36)

We are startled to find a universe we did not expect.

Literary Studies
Volume II (p. 403)

Barrow, John
In recent years cosmologists have begun to discuss the spontaneous creation of the Universe as a problem in physics. Those who do this assume that a future synthesis of quantum theory and relativity which reveals how gravity behaves when matter is enormously compressed will evade the predictions of a real singularity of the type required by the singularity theorems. Although the assumptions of the singularity theorems are not expected to hold near the singularity, we do not know whether to expect a singularity or not as yet. But even in the absence of this singularity to denote the beginning of the Universe, it has been speculated that the application of quantum theory to the whole Universe may allow physical content to be given to the concept of "creation of the Universe out of Nothing". The goal of this research is to show that the creation of an expanding universe is inevitable. The reason there is something rather than nothing is that "nothing" is unstable.

The World within the World (p. 230)

Bergson, Henri
The universe is not made, but is being made continually. It is growing, perhaps indefinitely . . .

Creative Evolution (p. 255)

Bloch, Arthur
The universe is simmering down, like a giant stew left to cook for four billion years. Sooner or later we won't be able to tell the carrots from the onions.

Quoted by John D. Barrow in
The World within the World (p. 221)

Bruno, Giordano
The center of the universe is everywhere, and the circumference nowhere.
Quoted by Joseph Silk in
The Big Bang (p. 84)

Burritt, Elijah H.
Beyond these are other suns, giving light and life to other systems, not a thousand, or two thousand merely, but multiplied without end, and ranged all around us, at immense distances from each other, attended by ten thousand times ten thousand worlds, all in rapid motion; yet calm, regular and harmonious—all space seems to be illuminated, and every particle of light a world . . . .. And yet all this vast assemblages of suns and worlds may bear no greater proportion to what lies beyond the utmost boundaries of human vision, than a drop of water to the ocean.
The Geography of the Heavens
Chapter XVI (p. 148)

Calder, Alexander
The universe is real but you can't see it. You have to imagine it.
Quoted by Katharine Kuh in
The Artist's Voice (p. 42)

Carlyle, Thomas
I don't pretend to understand the Universe—it's a great deal bigger than I am.
Quoted by D.A. Wilson and D.W. MacArthur in
Carlyle in Old Age (1865–1881) (p. 177)

Margaret Fuller: I accept the Universe.

Thomas Carlyle: Gad! she'd better!

Quoted by D.A. Wilson in
Carlyle on Cromwell and Others
Looking Round
Margaret Fuller Has to Listen (pp. 349–50)

Chekhov, Anton Pavlovich
But perhaps the universe is suspended on the tooth of some monster.
Note-Book of Anton Chekhov (p. 20)

Chesterton, Gilbert Keith
For the universe is a single jewel, and while it is a natural cant to talk of a jewel as peerless and priceless, of this jewel it is literally true. This

cosmos is indeed without peer and without price: for there cannot be another one.

Orthodoxy (pp. 116–7)

Clarke, Arthur C.
The universe: a device contrived for the perpetual astonishment of astronomers.

Quoted by Clifford A. Pickover in
Keys to Infinity (p. 41)

. . . the universes that drift like bubbles in the foam upon the River of Time.

Super Science Stories
The Wall of Darkness
Volume 5

Conger, George Perrigo
The universe as revealed in modern days and ways is so overwhelming that mind needs some other title than that of self-appointed legislator for it. Mind must register before it can regulate.

A World of Epitomizations
Introduction to Division Two (pp. 345–6)

Crane, Stephen
A man said to the universe:
"Sir, I exist!"
"However," replied the universe,
"The fact has not created in me
A sense of obligation."

The Collected Poems of Stephen Crane
Volume XXI
War is Kind (p. 101)

Crowley, Aleister
I have never grown out of the infantile belief that the universe was made for me to suck.

The Confessions of Aleister Crowley: an autohagiography
Chapter 54

d'Alembert, J.
To some one who could grasp the universe from a unified standpoint, the entire creation would appear as a unique truth and necessity.

Quoted by Charles W. Misner *et al*
Gravitation (p. 1218)

de Fontenelle, Bernard
. . . the universe is but a watch on a larger scale; all its motions depending on determined laws and mutual relation of its parts.

Conversations on the Plurality of Worlds
First Evening (p. 10)

Dee, John
The entire universe is like a lyre tuned by some excellent artificer, whose strings are separate species of the universal whole.

John Dee On Astronomy
XI

Donne, John
Man hath weaved out a net, and this net throwne upon the Heavens, and now they are his own . . .

Quoted by Roger A. MacGowan and Frederick I. Ordway, III in
Intelligence in the Universe
Ignatius His Conclave (p. 7)

Dyson, Freeman J.
As we look out into the Universe and identify the many accidents of physics and astronomy that have worked together to our benefit, it almost seems as if the Universe must in some sense have known that we were coming.

Quoted by John D. Barrow and Frank J. Tipler in
The Anthropic Cosmological Principle (p. 318)

Eddington, Sir Arthur Stanley
The unanimity with which the galaxies are running away looks almost as though they had a pointed aversion to us. We wonder why we should be shunned as though our system were a plague spot in the universe.

The Expanding Universe
The Recession of the Galaxies
Section III (p. 12)

I will give you a "celestial multiplication table." We start with a star as the unit most familiar to us, a globe comparable to the sun. Then—
A Hundred thousand million Stars make one Galaxy;
A hundred thousand million Galaxies
make one Universe.

The Expanding Universe
The Regression of the Galaxies
Section I (p. 4)

Eliot, T.S.
Do I dare
Disturb the universe?

> *The Complete Poems and Plays 1909–1950*
> The Love Song of J. Alfred Prufrock

Emerson, Ralph Waldo
Philosophically considered, the universe is composed of Nature and the Soul.

> *The Works of Ralph Waldo Emerson*
> Volume I
> Nature
> Introduction (p. 4)

Everything in the universe goes by indirection. There are no straight lines.

> *The Works of Ralph Waldo Emerson*
> Volume VII
> Society and Solitude
> Works and Days (p. 173)

Euler, Leonhard
For since the fabric of the universe is most perfect and the work of a most wise Creator, nothing at all takes place in the universe in which some rule of maximum or minimum does not appear.

> Quoted by Morris Kline in
> *Mathematical Thought from Ancient to Modern Times* (p. 573)

Field, Edward
Look, friend, at this universe
With its spiral clusters of stars
Flying out all over space
Like bedsprings suddenly bursting free;

> *New and Selected Poems from the Book of My Life*
> From STAND UP, FRIEND, WITH ME (1963)
> Prologue

Frost, Robert
The Universe is but the Thing of Things
The things but balls all going round in rings
Some of them mighty huge, some mighty tiny
All of them radiant and mighty shiny.

> *The Poetry of Robert Frost*
> Accidentally on Purpose
> l. 1–4

Because all reasoning is in a circle.
At least that's why the universe is round.

The Poetry of Robert Frost
Build the Soil
l. 202–3

Galilei, Galileo

No one will be able to read the great book of the Universe if he does not understand its language which is that of mathematics.

Quoted by A. Zee in
Fearful Symmetry (p. 122)

. . . I should think that anyone who considered it more reasonable for the whole universe to move in order to let the Earth remain fixed would be more irrational than one who should climb to the top of your cupola just to get a view of the city and its environs, and then demand that the whole countryside should revolve around him so that he would not have to take the trouble to turn his head.

Dialogue Concerning the Two Chief World Systems
The Second Day (p. 115)

Gamow, Barbara

"Your years of toil,"
Said Ryle to Hoyle,
   "Are wasted years, believe me.
The steady state
Is out of date.
   Unless my eyes deceive me,

My telescope
Has dashed your hope.
   Your tenets are uprooted.
Let me be terse:
Our universe
   Grows daily more diluted!"

Reproduced in George Gamow's
Matter, Earth and Sky (p. 596)

Genesis 1:1–4

In the beginning God created the heaven and the earth.

The Bible

Goethe, Johann Wolfgang von

Man is not born to solve the problems of the universe, but to find out where the problems begin, and then to take his stand within the limits of the intelligible.

Quoted by Louis Berman in
Exploring the Cosmos (p. 351)

Guth, Alan
The universe may be the ultimate free lunch.

Quoted by Herbert Friedman in
The Astronomer's Universe (p. 296)

Haldane, J.B.S.
. . . the universe is not only queerer than we suppose but it is queerer than we can suppose.

Possible Worlds and Other Papers
Possible Worlds (p. 298)

Hardy, G.H.
"Imaginary" universes are so much more beautiful than this stupidly constructed "real" one; and most of the finest products of an applied mathematician's fancy must be rejected, as soon as they have been created, for the brutal but sufficient reason that they do not fit the facts.

Mathematician's Apology
Section 26 (p. 135)

Harrison, Edward Robert
What determines the design of a universe; is it the Universe, God, fortuity, or the human mind?

Masks of the Universe (p. 5)

The universe consists only of atoms and the void: all else is opinion and illusion.

Masks of the Universe (p. 55)

Hawking, Stephen
I think the universe is completely self-contained. It doesn't have any beginning or end, it doesn't have any creation or destruction.

Quoted by Renée Weber in
Dialogues with Scientists and Sages (p. 89)

Even if there is only one possible unified theory, it is just a set of rules and equations. What is it that breathes fire into the equations and makes a universe for them to describe? The usual approach of science of constructing a mathematical model cannot answer the questions of why there should be a universe for the model to describe. Why does the universe go to all the bother of existing?

A Brief History of Time: From the Big Bang to Black Holes (p. 174)

There ought to be something very special about the boundary conditions of the universe and what can be more special than the condition that there is no boundary.

Quoted by John D. Barrow and Frank J. Tipler in
The Anthropic Cosmological Principle (p. 444)

Haynes, Margaret
The universe is just a bowl of spaghetti.

Quoted by Eric J. Lerner in
The Big Bang Never Happened (p. 49)

Heinlein, Robert A.
A zygote is a gamete's way of producing more gametes. This may be the purpose of the universe.

Time Enough for Love (p. 262)

Tomorrow I will seven eagles see, a great comet will appear, and voices will speak from whirlwinds foretelling monstrous and fearful things— This Universe never did make sense; I suspect that it was built on government contract.

The Number of the Beast (p. 14)

Holmes, John Haynes
The universe is not hostile, nor yet is it friendly. It is simply indifferent.

A Sensible Man's View of Religion
Is the Universe Friendly? (p. 39)

Hubble, Edwin
We find them smaller and fainter, in constantly increasing numbers, and we know that we are reaching out into space, further and ever further, until, with the faintest nebulae that can be detected with the greatest telescope, we arrive at the frontiers of the known universe.

Quoted by Joseph Silk in
The Big Bang (p. 26)

Jeans, Sir James Hopwood
. . . the Great Architect of the Universe now begins to appear as a pure mathematician.

The Mysterious Universe
Into the Deep Waters (p. 134)

The universe begins to look more like a great thought than a machine.

Quoted by Jefferson Hane Weaver in
The World of Physics
Volume II (p. 632)

The Universe can be best pictured, although still very imperfectly and inadequately, as consisting of pure thought, the thought of what, for want of a wider word, we must describe as a mathematical thinker.

The Mysterious Universe
Into the Deep Waters (p. 136)

Jeffers, Robinson
The universe expands and contracts like a great heart.
It is expanding, the farthest nebulae
Rush with the speed of light into empty space.

The Beginning and the End
The Great Explosion

Job 38:33
Knowest thou the ordinances of the heavens?

The Bible

Kepler, Johannes
This very cogitation carries with it I don't know what secret, hidden horror; indeed one finds oneself wandering in this immensity to which are denied limits and centre and therefore also all determinate places.

Quoted by Alexander Koyre in
From the Closed World to the Infinite Universe

Keyser, C.J.
Depend upon it, the universe will never really be understood unless it may be sometime resolved into an ordered multiplicity and made to own itself an everlasting drama of the calculus.

The Human Worth of Rigorous Thinking (p. 102)

Kunitz, Stanley
I see lines of your spectrum shifting red,
The Universe expanding, thinning out,
Our worlds flying, oh flying, fast apart.

Selected Poems 1928–1958
The Science of the Night

Kunz, F.L.
The whole universe is one mathematical and harmonic expression, made up of finite representations of the infinite.

Quoted by Renée Weber in
Dialogues with Scientists and Sages (p. 139)

Lederman, Leon
We hope to explain the entire universe in a single, simple formula that you can wear on your T-shirt.

Quoted in Richard Wolkomir's article
Quark City
Omni
February 1984 (p. 41)

Longfellow, Henry Wadsworth
The Universe, as an immeasurable wheel
Turning for evermore
In the rapid and rushing river of Time.
                              The Poetical Works of Henry Wadsworth Longfellow
                                                            Rain in Summer

Lucretius
. . . the existing universe is bounded in none of its dimensions; for then
it must have an outside.
                                                  On the Nature of Things
                                                        Book I, l. 958

McAleer, Neil
The known Universe "weighs" about
10,000,000,000,000,000,000,000,000,000,000,000,000,000,000,000 tons
—that is, ten trillion trillion trillion trillion tons.
                                              The Mind Boggling Universe (p. 165)

Marquis, Don
do not tell me
said warty bliggens
that there is not a purpose
in the universe
the thought is blasphemy
                                      the lives and times of archy & mehitabel
                                                  warty bliggens, the toad

the men of science are talking
about the size and shape of the universe again
I thought I had settled that for them
years ago it is as big as you think it is
and it is spherical in shape
                                      the lives and times of archy & mehitabel
                                                    why the earth is round

you write so many things
about me that are not true
complained the universe

there are so many things
about you which you seem to be
unconscious of yourself said archy

I contain a number of things
which I am trying to forget
rejoined the universe

such as what asked archy

such as cockroaches and poets
replied the universe

the lives and times of archy & mehitabel
poets

Milton, John

Defure [his] eyes in sudden view appear the secrets of the hoary Deep
dark illimitable ocean without bound without dimension . . .

Quoted by Roger A. MacGowan and Frederick I. Ordway, III in
Intelligence in the Universe (p. 7)

Hereafter, when they come to model Heav'n
And calculate the Starrs, how they will wield
The mightie frame, how build, unbuild, contrive
To save appeerances, how gird the Sphear
With Centric and Eccentric scribl'd o're,
Cycle and Epicycle, Orb in Orb.

Paradise Lost
Book VIII, l. 79–84

Morley, Christopher

The universe was dictated, but not signed.

Source unknown

Moulton, Forest Ray

It was pointed out by Lambert long ago that, just as the solar system is
a single unit in a galaxy of several hundred million stars, so the Galaxy
may be but a single one out of an enormous number of galaxies . . . and
that these galaxies may form larger units or supergalaxies, and so on
without limit.

Introduction to Astronomy (p. 469)

Noyes, Alfred

        "This universe
Exists, and by that one impossible fact
Declares itself a miracle; . . .

Watchers of the Sky
Newton I (p. 226)

Pagels, Heinz R.

. . . the universe contains the record of its past the way that sedimentary
layers of rock contain the geological record of the earth's past.

Perfect Symmetry (p. 24)

Pascal, Blaise
[The Universe] is an infinite sphere, the centre of which is everywhere, the circumference nowhere.

Pensées
Section II, 72

The spaces of the universe enfold me and swallow me up like a speck; but I, by the power of thought, may comprehend the universe.

Pensées
Section VI, 348

Penzias, Arno
Either we've seen the birth of the universe, or we've seen a pile of pigeon shit.

In Roylston Roberts
Serendipity

Phillpotts, Eden
The universe is full of magical things patiently waiting for our wits to grow sharper.

Quoted by John D. Barrow in
The World within the World (p. 300)

Pierce, C.S.
The universe ought to be presumed too vast to have any character.

Collected Papers
Volume VI

Pope, Alexander
He who thro' vast immensity can pierce,
See worlds on worlds compose one universe.
Observe how system into system runs,
What other planets circle other suns,
What varied being peoples every star,
May tell why Heav'n has made us as we are:

Complete Poetical Works of POPE
Essay on Man
Epistle I
Of the Nature and State of Man, with Respect to the Universe
Argument, l. 23–8

Order is Heav'n's first law.

Complete Poetical Works of POPE
An Essay on Man
Epistle IV, l. 49

Prigogine, Ilya
I certainly think we are only living in the prehistory of the understanding of our universe.

Quoted by Renée Weber in
Dialogues with Scientists and Sages (p. 199)

Rabi, I.I.
The scientist does not defy the universe. He accepts it. It is his dish to savor, his realm to explore; it is his adventure and never-ending delight. It is complaisant and elusive but never dull. It is wonderful both in the small and in the large. In short, its exploration is the highest occupation for a gentleman.

Quoted by Leon Lederman in
The God Particle (p. 104)

Raymo, Chet
Give me the ninety-two elements and I'll give you a universe. Ubiquitous hydrogen. Standoffish helium, Spooky boron. No-nonsense carbon. Promiscuous oxygen. Faithful iron. Mysterious phosphorous. Exotic xenon. Brash tin. Slippery mercury. Heavy-footed lead.

The Soul of the Night (p. 65)

Reed, Ishmael
The universe is a spiraling Big Band in a polka-dotted speak-easy, effectively generating new lights every one-night stand.

Quoted by A. Zee in
An Old Man's Toy (p. 123)

Rindler, Wolfgang
Modern scientific man has largely lost his sense of awe in the Universe. He is confident that given sufficient intelligence, perseverance, time, and money, he can understand all there is beyond the stars.

Quoted by M. Taube in
Evolution of Matter and Energy (p. 18)

Ruderman, M.A.
Rosenfeld, A.H.
We are peeling an onion layer by layer, each layer uncovering in a sense another universe, unexpected, complicated, and—as we understand more—strangely beautiful.

American Scientist
An Explanatory Statement on Elementary Particle Physics (p. 210)
Volume 48, Number 2, June 1960

Russell, Bertrand

All the labors of the ages, all the devotion, all the inspiration, all the noonday brightness of human genius, are destined to extinction in the vast death of the solar system; and the whole temple of Man's achievement must inevitably be buried beneath the debris of a universe in ruins.

Quoted by George Smoot in
Wrinkles in Time (p. 69)

Sagan, Carl

The Cosmos is all that is or ever was or ever will be.

Cosmos (p. 4)

"To my mind, it seems not fully satisfactory to say that there was a first cause. That seems to postpone dealing with the problem rather than solving it. If we say 'God' made the universe, then surely the next question is 'Who made God?' If we say 'God was always here', why not say the universe was always here? . . . "

US Catholic
May 1981 (p. 20)

Sandage, Allan

The present universe is something like the old professor nearing retirement with his brilliant future behind him.

Quoted by G. Borner in
The Early Universe (p. 90)

Scientific American Editorial

It is obvious that we must regard the universe as extending infinitely, forever, in every direction; or that we must regard it as not so extending. Both possibilities go beyond us.

Scientific American
Einstein's Finite Universe (p. 202)
Volume CXXIV, Number 11, March 12, 1921

Shaw, George Bernard

As an Englishman [Newton] postulated a rectangular universe because the English always use the word "square" to denote honesty, truthfulness, in short: rectitude. Newton knew that the universe consisted of bodies in motion, and that none of them moved in straight lines, nor ever could. But an Englishman was not daunted by the facts. To explain why all the lines in his rectilinear universe were bent, he invented a force called gravitation and then erected a complex British universe and established it as a religion which was devoutly believed for 300 years. The book of this Newtonian religion was not that oriental magic thing, the Bible. It was that British and matter-of-fact thing, a Bradshaw

[English railway timetable]. It gives the stations of all the heavenly bodies, their distances, the rates at which they are traveling, and the hour at which they reach eclipsing points or crash into the earth. Every item is precise, ascertained, absolute and English. Three hundred years after its establishment a young professor rises calmly in the middle of Europe and says to our astronomers: "gentlemen: if you will observe the next eclipse of the sun carefully, you will be able to explain what is wrong with the perihelion of Mercury." . . . The young professor smiles and says that gravitation is a very useful hypothesis and gives fairly close results in most cases, but that personally he can do without it. He is asked to explain how, if there is no gravitation, the heavenly bodies do not move in straight lines and run clear out of the universe. He replies that no explanation is necessary because the universe is not rectilinear and exclusively British; it is curvilinear. The Newtonian Universe thereupon drops dead and is supplanted by the Einstein Universe. Einstein has not challenged the facts of science but the axioms of science, and science has surrendered to the challenge.

> Quoted by John D. Barrow in
> *The World within the World* (p. 109)

Shelley, Percy Bysshe
Below lay stretched the boundless universe!
  There, far as the remotest line
That limits swift imagination's flight,
Unending orbs mingled in mazy motion,
  Immutably fulfilling
  Eternal Nature's law.
  Above, below, around,
  The circling systems formed
  A wilderness of harmony—
  Each with undeviating aim
In eloquent silence through the depths of space
  Pursued its wondrous way.

> *The Complete Poetical Works of Shelley*
> The Daemon of the World
> Part I, l. 241–52

Silk, Joseph
The development of human awareness of the Universe evolved from the geocentric cosmology of the ancient world via the heliocentric cosmology of the Renaissance and the egocentric cosmology of the nineteenth century, to the ultimate destination of the Big-Bang theory of the expanding Universe.

> *Cosmic Enigmas* (p. 3)

Smith, Logan
I woke this morning . . . into the well-known, often-discussed, but, to my mind, as yet unexplained Universe.

Trivia
To-Day (p. 4)

Trimble, V.
Those of us who are not directly involved in the fray can only suppose that the universe is open ($\Omega < 1$) on Wednesday, Friday, and Sunday and closed ($\Omega > 1$) on Thursday, Saturday, and Monday. (Tuesday is choir practice.)

Contemporary Physics
Dark Matter in the Universe: Where, What, and Why? (p. 389)
Volume 29, 1988

Unknown
We will first understand
How simple the universe is
When we realize
How strange it is.

Source unknown

Astronomers say the universe is finite, which is a comforting thought for those people who can't remember where they leave things.

Source unknown

Weinberg, Steven
It is very hard to realize that this all is just a tiny part of an overwhelmingly hostile universe. It is even harder to realize that this present universe has evolved from an unspeakably unfamiliar early condition, and faces a further extinction of endless cold or intolerable heat. The more the universe seems comprehensible, the more it also seems pointless.

The First Three Minutes
Epilogue (p. 154)

The effort to understand the universe is one of the very few things that lifts human life a little above the level of farce, and gives it some of the grace of tragedy.

The First Three Minutes
Epilogue (p. 155)

Whitman, Walt
The world, the race, the soul—
  Space and time, the universes
All bound as is befitting each—all
Surely going somewhere.

Complete Poems and Collected Prose
Going Somewhere

Zebrowski, George
The rationality of our universe is best suggested by the fact that we can discover more about it from any starting point, as if it were a fabric that will unravel from any thread.

OMNI
Is Science Rational? (p. 50)
June 1994

In a perfectly rational universe, infinities turn back on themselves . . .

OMNI
Is Science Rational? (p. 50)
June 1994

SAINT AUGUSTINE ERA

Cardenal, Ernesto
Before the big explosion
  there wasn't even empty space,
    for space and time, and matter and energy, emerged from the
      explosion,
  neither was there any "outside" into which the universe could explode
for the universe embraced it all, even the whole of empty space.

Cosmic Canticle
Cantiga 1
Big Bang

Saint Augustine
See, I answer him that asketh, "What did God before He made heaven and earth?" I answer not as one is said to have done merrily (eluding the pressure of the question): "He was preparing hell (saith he) for pryers into mysteries."

The Confessions
Book XI, XII, 14

Tzu, Lao
Something mysteriously formed,
Born before heaven and earth.
In the silence and the void,

Standing alone and unchanging,
Ever present and in motion.

Tao Te Ching
Twenty-five

COSMOGENESIS

Ackerman, Diane
Fifteen billion years ago, when the Universe
let rip and, in disciplined panic,
Creation
spewed
mazy star-treacle and resin,
shrinking balls of debut fire smoldered
and
            glitched.

The Planets: A Cosmic Pastoral
Neptune
IV

Adams, Douglas
In the beginning the universe was created. This made a lot of people very
angry and been widely regarded as a bad move. Many races believe that
it was created by some sort of god, though the Jatravartid people of
Viltvodle VI believe that the entire Universe was in fact sneezed out of
the nose of a being called the Great Green Arkleseizure.

The Restaurant at the End of the Universe
Chapter 1

Cardenal, Ernesto
In the beginning there was nothing
  neither space
    nor time.
      The entire universe concentrated
in the space of the nucleus of an atom,
and before that even less, much less than a proton,
and even less still, an infinitely dense mathematical point.

Cosmic Canticle
Cantigua 1
Big Bang

Eddington, Sir Arthur Stanley
Since I cannot avoid introducing this question of beginning, it has seemed
to me that the most satisfactory theory would be one which made the
beginning *not too unaesthetically abrupt.*

The Expanding Universe (p. 56)

Egyptian Myth c. 2500 B.C.
In the beginning, only the ocean existed, upon which there appeared an egg. Out of the egg came the sun-god and from himself he begat four children: Shu and Tefnut, Keb and Nut. All these, with their father, lay upon the ocean of chaos. Then Shu and Tefnut thrust themselves between Keb and Nut. They planted their feet upon Keb and raised Nut on high so that Keb became the earth and Nut the heavens.

Quoted by Eric J. Lerner in
The Big Bang Never Happened (p. 58)

Flaubert, Gustave
SMARH: How vast creation is! I see the planets rise, I see the fiery stars driven along . . ... Space opens out as I rise, worlds revolve around me, and I am the center of this bustling creation.

Early Writings
Smarh (p. 216)

Gamow, George
In the beginning God created radiation and ylem. And ylem was without shape or number, and the nucleons were rushing madly over the face of the deep.

My World Line (p. 127)

Genesis 1:1
In the beginning God created the heaven and the earth.

The Bible

Hein, Robert
The first world the cosmic colossus created with a word:
One lightning word from the golden lips of Truth
And electric earth condensed on a creamy cloud,
Adorned with a necklace of blue–gold stars and a chain
Of peppermint planets in the amphitheater of space . . .

Quest of the Singing Tree
The Larger Creation
Creation of the Earth

Hoyle, Fred
Without continuous creation, the Universe must evolve toward a dead state in which all the matter is condensed into a vast number of dead stars.

The Nature of the Universe
The Expanding Universe (pp. 131–2)

Pima Creation Myth
In the beginning there was nothing at all except darkness . . ... The Earth Doctor saw that when the sun and moon were not in the sky, all was

inky darkness. So he sang a magic song, and took some water into his mouth and blew it into the sky, in a spray, to make little stars . . .. Next he took his walking stick and placed ashes on the end of it. Then he drew it across the sky to form the Milky Way . . .

Quoted by Raymond Van Over in
Sun Songs (p. 28)

Sagan, Carl

Na Arean sat alone in space as a cloud that floats in nothingness. He slept not, for there was no sleep; he hungered not, for as yet there was no hunger. So he remained for a great while, until a thought came to his mind. He said to himself, "I will make a thing."

Cosmos (p. 25)

Saint Augustine

And aught else besides Thee was there not, whereof Thou mightest create them, O God, One Trinity, and Trine Unity; and therefore out of nothing didst Thou create heaven and earth; a great thing, and a small thing; for Thou art Almighty and Good, to make all things good, even the greatest heaven and the petty earth. Thou wert, and nothing was there besides, out of which Thou createdst heaven and earth; things of two sorts: one near Thee, the other near to nothing; one to which Thou alone shouldest be superior, the other to which nothing should be inferior.

Confessions
XII, 7

Singer, Isaac Bashevis

"Who created the world?"

"There was matter somewhere in the cosmos and for a long time it lay there and stank. That stench was the origin of life."

"Where did the matter come from?"

"What is the difference? The main thing is that we have no responsibility—neither to ourselves nor to others. The secret of the universe is apathy. The earth, the sun, the rocks, they're all indifferent, and this is a kind of passive force. Perhaps indifference and gravitation are the same."

He spoke and yawned. He ate and smoked.

"Why do you smoke so much?" I asked.

"It keeps me indifferent."

A Crown of Feathers
The Captive (p. 47)

Singh, Jagjit
In the beginning there was neither heaven nor earth,

And there was neither space nor time.

And the Earth, the Sun, the Stars, the Galaxies and the whole universe were confined within a small volume like the bottled genie of the Arabian Nights.

And then God said, "Go!"

And straight way the Galaxies rushed out of their prison, scattering in all directions, and they have continued to run away from one another ever since, afraid lest some cosmic Hand should gather them again and put them back in the bottle (which is not bigger than a pin-point).

And they shall continue to scatter thus till they fade from each other's ken —and thus, for each other, cease to exist at all.

Great Ideas of Modern Mathematics
Space and Time (pp. 209–10)

Spenser, Edmund
Through knowledge we behold the world's creation,
How in his cradle first he fostered was;
And judge of Natures cunning operation,
How things she formed of a formless mass . . .

The Complete Poetical Works of Edmund Spenser
The Tears of the Muses
l. 499–502

Sufi Creation Myth
I was a hidden treasure and desired to be known: therefore I created the creation in order to be known.

Quoted by George Smoot in
Wrinkles in Time (p. 1)

Tagore, Rabindranath
It seems to me that, perhaps, creation is not fettered by rules,
That all the hubbub, meeting and mingling are
    blind happenings of fate . . .

Our Universe (p. 75)

Updike, John
By computation, they all must have begun at one place about five billion years ago; all the billions and trillions and quadrillions squared and squared again of tons of matter in the universe were compressed into a ball at the maximum possible density, the density within the nucleus

of the atom; one cubic centimeter of this primeval egg weighed two
hundred and fifty tons.

The Centaur (p. 38)

Weinberg, Steven

. . . the urge to trace the history of the universe back to the beginnings
is irresistible. From the start of modern science in the sixteenth and
seventeenth centuries, physicists and astronomers have returned again
and again to the problem of the origin of the universe.

The First Three Minutes
Chapter I (p. 4)

Wilmot, John [Rochester, John Wilmot Earl of]

Eer time and place were, time and place were not,
When Primitive Nothing something straight begot,
Then all proceeded from the great united—What.

Collected Works of John Wilmot Earl of Rochester
Upon Nothing

Zuni Creation Myth

In the beginning of things Awonawilona was alone. There was nothing
beside him in the whole of time. Everywhere there was black darkness
and void. Then Awonawilona conceived in himself the thought, and the
thought took shape and got out into space and through this stepped out
into the void, into outer space, and from them came nebulae of growths
and mists, full of power and growth.

Quoted by Raymond Van Over in
Sun Songs (p. 23)

DYING

Byron, Lord George Gordon

I had a dream, which was not all a dream.
The bright sun was extinguish'd, and the stars
Did wander darkling in the eternal space,
Rayless, and pathless, and the icy earth
Swung blind and blackening in the moonless air . . .

The Poetical Works of Lord Byron
Miscellaneous Poems
Darkness

Dyson, Freeman J.

Since the universe is on a one-way slide toward a state of final death
in which energy is maximally degraded, how does it manage, like King
Charles, to take such an unconsciously long time a-dying.

Quoted by John D. Barrow and Frank J. Tipler in
The Anthropic Cosmological Principle (p. 385)

Eliot, T.S.
This is the way the world ends
Not with a bang but a whimper.

The Collected Poems and Plays 1909–1950
The Hollow Men
v

Frost, Robert
Some say the world will end in fire,
Some say in ice.
From what I've tasted of desire
I hold with those who favor fire.
But if it had to perish twice,
I think I know enough of hate
To say that for destruction ice
Is also great
And would suffice.

The Poetry of Robert Frost
Fire and Ice

Gamow, George
Galaxies are ever spinnik,
 Stars will burn to final sparrk,
Till ourr universe is thinnink
 And is lifeless, cold and darrk.

Mr. Tompkins in Paperback
Chapter 6 (p. 60)

Gribbin, John
"Big Crunch" is an ugly term which hardly seems appropriate for so important an event as the end of the universe. But there is no convention as yet for a label of the moment of destruction at the end of time, and I am free to borrow the term 'omega point.'

The Omega Point: The Search for the Missing Mass
and the Ultimate Fate of the Universe (p. 2)

Indian Myth
One thousand mahāyugas—4,320,000,000 years of human reckoning—constitute a single day of Brahmā, a single kalpa . . . It endures for only one hundred Brahmā years of Brahmā days and nights, and concludes with a great, or universal, dissolution. Then vanish not only the visible spheres of the three worlds (earth, heaven and the space between), but all spheres of being whatsoever, even those of the highest worlds. All become resolved into the divine, primeval substance. A state of total

re-absorption then prevails for another Brahmā century, after which the entire cycle of 311,040,000,000,000 human years begins anew.

Quoted by H. Zimmer in
Myths and Symbols in Indian Art and Civilization (p. 18.9)

Jeffers, Robinson

It will contract, the immense navies of stars and galaxies,
  dust-clouds and nebulae
Are recalled home, they crush against each other in one
  harbor, they stick in one lump . . .

The Beginning and the End
The Great Explosion

Lucretius

And so some day,
The mighty ramparts of the mighty universe
Ringed round with hostile force,
Will yield and face decay and come crumbling to ruin.

Quoted by Paul Davies in
About Time (p. 33)

Nicholson, Norman

And if the universe
Reversed and showed
The colour of its money;
If now observable light
Flowed inward, and the skies snowed
A blizzard of galaxies,
The lens of night would burn
Brighter than the focussed sun,
And man turn blinded
With white-hot darkness in his eyes.

The Pot Geranium
The Expanding Universe

Russell, Bertrand

In the vast death of the solar system, the whole temple of Man's achievements must inevitably be buried beneath all the labours of the ages, all the devotion, all the inspiration, all the noonday brightness of human genius, are destined to extinction debris of a universe in ruins— all these things, if not quite beyond dispute, are yet so nearly certain that no philosophy which rejects them can hope to stand. Only within the scaffolding of these truths, only on the firm foundation of unyielding despair, can the soul's habitation henceforth be safely built.

Why I am not a Christian (p. 107)

VACUUM

Marquis, Don
he i said is afraid of a vacuum what is
there in a vacuum to make one afraid
said the flea there is nothing in it
I said and that is what makes one
afraid to contemplate it a person
can t think of a place with nothing at
all in it without going nutty and if he
tries to think that nothing is
something after all he gets nuttier . . .

> *the lives and times of archy & mehitabel*
> the merry flea

Pascal, Blaise
Because . . . you have believed from childhood that a box was empty
when you saw nothing in it, you have believed in the possibility of a
vacuum.

> *Pensées*
> Section II, 82

. . . I had always held that the vacuum was not a thing impossible in
nature and that she did not flee it with such horror as many imagine.

> *Treatise Concerning the Vacuum*
> New Experiments Concerning the Vacuum

Whitehead, Alfred North
You cannot have first space and then things to put into it, any more than
you can have first a grin and then a Cheshire cat to fit on to it.

> Quoted by Sir Arthur Eddington in
> *New Pathways in Science* (p. 48)

Williams, Tennessee
. . . a vacuum is a hell of a lot better than some of the stuff that nature
replaces it with.

Cat on a Hot Tin Roof
Act Two

VELOCITY

Unknown

Velocity is what you let go of a hot frying-pan handle with.

Source unknown

VOID

Aristotle
Let us explain . . . that there is no void existing separately, as some maintain.

Physics
214b12

Bruno, Giordano
There is a single general space, a single vast immensity which we may freely call Void; in it are innumerable globes like this on which we live and grow. This space we declare to be infinite; since neither reason, convenience, possibility, sense-perception nor nature assign to it a limit. In it are an infinity of worlds of the same kind as our own. For there is no reason nor defect of nature's gifts, either of active or of passive power, to hinder the existence of other worlds throughout space, which is identical in natural character with our own space.

De l'Infinito Universo et Mondi
Quoted by Eli Maor in
To Infinity and Beyond: A Cultural History of the Infinite (p. 198)

Pagels, Heinz R.
The nothingness "before" the creation of the universe is the most complete void that we can imagine—no space, time, or matter existed. It is a world without place, without duration or eternity, without number— it is what mathematicians call "the empty set." Yet this unthinkable void converts itself into the plenum of existence—a necessary consequence of physical laws. Where are these laws written into that void? What "tells" the void that it is pregnant with a possible universe? It would seem that even the void is subject to law, a logic that exists prior to space and time.

Perfect Symmetry (p. 347)

Smith, Logan
. . . I cool my thoughts with a vision of the giddy, infinite, meaningless waste of Creation, the blazing Suns, the Planets and frozen Moons, all crashing blindly forever across the void of space.

Trivia
Mental Vice (p. 97)

Thomson, James
With what an awful,
  world-revolving power,
Were first the unwieldy
  planets launched along
The illimitable void! There
  to remain
Amidst the flux of many
  thousand years,
That oft has swept the
  toiling race of men,
And all their labored
  monuments, away.

Quoted by Eli Maor in
To Infinity and Beyond: A Cultural History of the Infinite (p. 206)

X-RAY

Jauncey, G.E.M.
O Roentgen, then the news is true
  And not a trick of idle rumor
That bids us each beware of you
  And of your grim and graveyard humor.

Scientific American
February 22, 1896

Polanyi, Michael
One of the greatest and most surprising discoveries of our own age, that of the diffraction of X-rays by crystals (in 1912) was made by a mathematician, Max von Laue, by the sheer power of believing more concretely than anyone else in the accepted theory of crystals and X-rays.

Personal Knowledge (p. 277)

Russell, Bertrand
Everybody knows something about X-rays, because of their use in medicine. Everybody knows that they can take a photograph of the skeleton of a living person, and show the exact position of a bullet lodged in the brain. But not everybody knows why this is so. The reason is that the capacity of ordinary matter for stopping the rays varies approximately as the fourth power of the atomic number of the elements concerned . . .

The ABC of Atoms (p. 106)

Russell, L.K.
She is so tall, so slender; and her bones—
  Those frail phosphates, those carbonates of lime,—
  Are well produced by cathode rays sublime,
By oscillations, amperes and by ohms.

Her dorsal vertebrae are not concealed
By epidermis, but are well revealed.

Life
Line on an X-ray Portrait of a Lady (p. 191)
Volume XXVII, Number 689, March 12, 1896

Unknown
The Roentgen Rays, The Roentgen Rays
What is this craze,
The town's ablaze,
With the new phase
of X-rays' ways
I'm full of daze,
Shock and amaze,
For nowadays,
I hear they'll gaze,
Thro' cloak and gown—and even stays,
These naughty, naughty Roentgen Rays.

Quoted by John G. Taylor in
The New Physics (p. 46)

BIBLIOGRAPHY

Ackerman, Diane. *The Planets: A Cosmic Pastoral*. William Morrow and Company, Inc., New York. 1976.

Adair, Robert K. *The Great Design*. Oxford University Press, Inc., New York. 1987.

Adams, Douglas. *Life, the Universe and Everything*. Harmony Books, New York. 1982.

Adams, Douglas. *Mostly Harmless*. Harmony Books, New York. 1992.

Adams, Douglas. *The Restaurant at the End of the Universe*. Harmony Books, New York. 1980.

Adams, Henry. *The Education of Henry Adams*. Random House, New York. 1946.

Aeschylus. *Agamemnon* in *Great Books of the Western World*. Volume 5. Encyclopædia Britannica, Inc., Chicago. 1952.

Alighieri, Dante. *Paradiso*. J.M. Dent, London. 1962.

Alighieri, Dante. *The Divine Comedy of Dante Alighiere* in *Great Books of the Western World*. Translated by Charles Eliot Norton. Volume 21. Encyclopædia Britannica, Inc., Chicago. 1952.

Allen, Richard Hinckley. *Star Names: Their Lore and Meaning*. G.E. Stretcher, New York. 1899.

Allen, Woody. *Getting Even*. Random House, New York. 1971.

Anscombe, F.J. 'Rejection of Outliers' in *Technometrics*. Volume 2, Number 2. May 1960.

Archimedes. *The Works of Archimedes* translated by Sir Thomas L. Heath. Cambridge University Press. 1897.

Aristotle. *Categories* in *Great Books of the Western World*. Translated by E.M. Edghill. Volume 8. Encyclopædia Britannica, Inc., Chicago. 1952.

Aristotle. *Meteorology* in *Great Books of the Western World*. Translated by E.W. Webster. Volume 8. Encyclopædia Britannica, Inc., Chicago. 1952.

Aristotle. *On Generation and Corruption* in *Great Books of the Western World*. Translated by H.H. Joachim. Volume 8. Encyclopædia Britannica, Inc., Chicago. 1952.

Aristotle. *On the Heavens* in *Great Books of the Western World*. Translated by J.L. Stocks. Volume 8. Encyclopædia Britannica, Inc., Chicago. 1952.

Aristotle. *Physics* in *Great Books of the Western World*. Translated by R.P. Hardie and R.K. Gaye. Volume 8. Encyclopædia Britannica, Inc., Chicago. 1952.

Aristotle. *Politics* in *Great Books of the Western World*. Translated by Benjamin Jowett. Volume 9. Encyclopædia Britannica, Inc., Chicago. 1952.

Aristotle. *Posterior Analytics* in *Great Books of the Western World*. Translated by G.R.G. Mure. Volume 8. Encyclopædia Britannica, Inc., Chicago. 1952.

Aurelius, Marcus. *The Meditations of Marcus Aurelius*. Walter Scott, London. 1887.

Aurobindo. *The Riddle of the World*. Arya Publishing House, Calcutta. 1946.

Bacon, Francis. *Advancement of Learning* in *Great Books of the Western World*. Volume 30. Encyclopædia Britannica, Inc., Chicago. 1952.

Bacon, Francis. *Novum Organum* in *Great Books of the Western World*. Volume 30. Encyclopædia Britannica, Inc., Chicago. 1952.

Bacon, Leonard. *Rhyme and Punishment*. Rinehart & Company, Inc., New York. 1936.

Baez, Joan. *Daybreak*. Avon Books, New York. 1968.

Bagehot, Walter. *Literary Studies*. Volumes I and II. Dent, London. 1911.

Bailey, Philip James. *Festus*. George Routledge and Sons, Limited. London. 1893.

Ball, Walter William Rouse. *A Short Account of the History of Mathematics*. Macmillan, London. 1908.

Barnett, Lincoln. *The Universe and Dr. Einstein*. William Sloane Associates, New York. 1948.

Barrow, John D. *The Artful Universe*. Clarendon Press, Oxford. 1995.

Barrow, John D. *The World within the World*. Clarendon Press, Oxford. 1988.

Barrow, John D. *Theories of Everything*. Clarendon Press, Oxford. 1991.

Barrow, John D. and Tipler, Frank J. *The Anthropic Cosmological Principle*. Oxford University Press, Oxford. 1986.

Barry, Frederick. *The Scientific Habit of Thought*. Columbia University Press, New York. 1927.

Bartusiak, Marcia. *Thursday's Universe*. Times Books, New York. 1986.

Baumel, Judith. *The Weight of Numbers*. Wesleyan University Press, Middleton. 1988.

Beerbohm, Max. *Mainly on the Air*. Alfred A. Knopf, New York. 1958.

Beiser, Arthur. *The World of Physics*. McGraw-Hill Book Company, Inc., New York. 1960.

Bell, E.T. *Debunking Science*. University of Washington Book Store, Seattle. 1930.

Bell, E.T. *Men of Mathematics*. Simon and Schuster, New York. 1965.

Bentley, Richard. *The Works of Richard Bentley*. Volume III. Francis Macpherson, London. 1838.

Berger, A. *The Big Bang and Georges Lemaître*. Dordrecht, Boston. 1984.

Bergerac, Cyrano. *The Comical History of the States and Empires of the Worlds of the Moon and Sun*. Translated by A. Lovell. Printed for *Henry Rhodes*, next door to the *Swan Tavern*, near *Bride Lane*, in *Fleet Street*, London. 1687.

Bergson, Henri. *Creative Evolution*. Henry Holt and Company, New York. 1911.

Berkeley, Edmund C. 'Right Answers–A Short Guide for Obtaining Them' in *Computers and Automation*. Volume 18, Number 10. September 1969.

Berkeley, George. *The Principles of Human Knowledge*. P. Smith, Glouster. 1978.

Berman, Louis. *Exploring the Cosmos*. Little, Brown and Company, Boston. 1973.

Bernard, Claude. *An Introduction to the Study of Experimental Medicine*. H. Schuman, New York. 1927.

Bernstein, Jeremy. *Experiencing Science*. Basic Books Inc., New York. 1978.

Beveridge, W.I.B. *The Art of Scientific Investigation*. Norton, New York. 1957.

Bierce, Ambrose. *The Enlarged Devil's Dictionary*. Doubleday & Company, Inc., Garden City. 1967.

Billings, Josh. *Old Probability: Perhaps Rain—Perhaps Not*. G.W. Carlton and Co., New York. 1879.

Birkhoff, George David. 'Mathematical Nature of Physical Theories' in *American Scientist*. Volume 31, Number 4. October 1943.

Birks, J.B. *Rutherford at Manchester*. W.A. Benjamin Inc., New York. 1963.

Bishop, Morris. *A Bowl of Bishop*. Dial Press, New York. 1954.

Blackwood, Oswald. *Introductory College Physics*. John Wiley & Sons, Inc., New York. 1939.

Blake, William. *The Prophetic Writings of William Blake*. At the Clarendon Press, Oxford. 1926.

Blake, William. *BLAKE: The Complete Poems*. Longman, London. 1989.

Blake, William. *The Letters of William Blake*. Edited by Geoffrey Keynes. Rupert HART-DAVIS, London. 1956.

Blavatsky, H.P. *Isis Unveiled*. Theosophy Co., Los Angeles. 1968.

Boas, Ralph P., Jr. *Lion Hunting & Other Mathematical Pursuits*. Mathematical Association of America, Washington, D.C. 1995.

Bohr, Niels. *Atomic Theory and the Description of Nature*. At the University Press, Cambridge. 1961.

Boodin, John Elof. *Cosmic Evolution*. The Macmillan Company, New York. 1925.

Borel, Émile. *Probability and Certainty*. Walker, New York. 1963.

Borel, Émile. *Probabilities and Life*. Translated by Maurice Baudin. Dover Publications, New York. 1962.

Borges, Jorge Luis. *A Personal Anthology*. J. Cape, London. 1968.

Borges, Jorge Luis. *Cuentos de Jorge Luis Borges*. The Monticello College Press, Godfrey. 1958.

Borges, Jorge Luis. *Ficciones*. Grove Press, New York. 1962.

Born, Max. *Experiment and Theory in Physics*. At the University Press, Cambridge. 1944.

Born, Max. *Natural Philosophy of Cause and Chance*. Dover Publications, New York. 1964.

Born, Max. *The Born–Einstein Letters*. Walker and Company, New York. 1971.

Born, Max. *The Restless Universe*. Dover Publications, New York. 1951.

Borner, G. *The Early Universe*. Springer-Verlag, Berlin. 1988.

Boswell, James. *Boswell's Life of Johnson*. Oxford University Press, London. 1927.

Boyle, Robert. *Usefulness of Mathematics to Natural Philosophy*. Printed for A. Millary, London. 1744.

Box, G.E.P. 'Discussion' in *Journal of the Royal Statistical Society*. Series B, Volume 18. 1956.

Box, G.E.P. 'Use and Abuse of Regression' in *Technometrics*. Volume 8, Number 4. November 1966.

Bradley, Omar. *The Collected Writings of General Omar N. Bradley*. Washington. 1967–71.

Bragg, W. *The Universe of Light*. Macmillan, New York. 1931.

Bragg, W. 'Electrons and Ether Waves' in *Scientific Monthly*. Volume 14. February 1922.

Brecht, Bertolt. *Galileo*. Translated by Charles Laughton. Grove Weidenfeld, New York. 1966.

Brewster, Sir David. *Memoirs of the Life, Writings, and Discoveries of Sir Isaac Newton*. Volume 2. Edinburgh, Edmonston. 1860.

Bridgman, P.W. *The Logic of Modern Physics*. The Macmillan Company, New York. 1927.

Bridgeman, P.W. *The Nature of Physical Theory*. Princeton University Press, Princeton. 1936.

Bronowski, Jacob. *The Ascent of Man*. Little Brown, Boston. 1973.

Brood, William J. 'Golden Age of Astronomy Peers to the Edge of the Universe' in *NY Times*. May 8, 1984.

Brousson, Jean Jacques. *Anatole France Himself*. J.B. Lippincott Company, Philadelphia. 1925.

Brown, John. *Horae Subsecivae*. Oxford University Press, London. 1907.

Browne, Sir James Crichton. *From the Doctor's Notebook*. Duckworth, London. 1937.

Browning, Robert. *The Poems of Robert Browning*. Heritage Press, New York. 1971.

Bruncken, Herbert Gerhardt. 'A Space–Time Lullaby' in *The Physics Teacher*. April 1963.

Bruner, Jerome Seymore. *The Process of Education*. Harvard University Press, Cambridge. 1960.

Bryant, Edward. 'Winslow Crater' in *Fantasy and Science Fiction*. 1980.

Buck, Pearl S. 'The Bomb—The End of the World?' in *American Weekly*. March 8, 1959.

Buller, Arthur Henry Reginald. 'Relativity' in *Punch*. Volume CLXV. December 19, 1923.

Bunting, Basil. *The Complete Poems*. Oxford University Press, Oxford. 1978.

Burgel, Bruno H. *Astronomy for All*. Cassell and Company, Ltd., London. 1911.

Burritt, Elijah H. *The Geography of the Heavens*. Mason Brothers, Boston. 1863.

Burroughs, William. *The Adding Machine*. Seaver Books, New York. 1986.

Burton, David M. *The History of Mathematics*. Wm. C. Brown Publishers, Dubuque. 1988.

Bush, Douglas. *Science and English Poetry*. Oxford University Press, New York. 1950.

Butler, Samuel. *Samuel Butler's Notebooks*. Edited by Geoffrey Keyner and Brian Hill. Jonathan Cape, London. 1951.

Byron, Lord. *Childe Harold's Pilgrimage*. Cassell, London. 1886.

Byron, Lord. *The Poetical Works of Lord Byron*. Jas. B. Smith & Company, Philadelphia. 1859.

Cajori, Florian. *Sir Isaac Newton's Mathematical Principles of Natural Philosophy and His System of the World*. University of California Press, Berkeley. 1934.

Cajori, Florian. *The Teaching and History of Mathematics in the United States*. Government Printing Office, Washington D.C. 1890.

Calder, Ritchie. *Profile of Science*. Allen & Unwin, London. 1951.

Campbell, Lewis and Garnett, William. *The Life of James Clerk Maxwell*. Macmillan and Company, London. 1882.

Camus, Albert. *The Myth of Sisyphus*. Translated from the French by Justin O'Brien. Vintage Books, New York. 1955.

Camus, Albert. *The Fall*. Vintage Books, New York. 1956.

Čapek, Milič. *The Philosophical Impact of Contemporary Physics*. Van Nostrand, New Jersey. 1961.

Cardenal, Ernesto. *Cosmic Canticle*. Curbstone Press, Willimantic. 1993.

Carlyle, Thomas. *Sartor Resartus*. G. Bell & Sons, London. 1909.

Carnap, Rudolf. *The Unity of Science*. Translated by M. Black. Thoemmes Press, Bristol. 1995.

Carroll, Lewis. *The Complete Works of Lewis Carroll*. Modern Library, New York. 1936.

Carus, Paul. 'The Nature of Logical and Mathematical Thought' in *Monist*. Volume 20, Number 1. 1910.

Cedering, Siv. *Letters from the Floating World*. University of Pittsburg Press, Pittsburg. 1984.

Cerf, Bennett. *Out on a Limerick*. Harper and Bros., New York. 1960.

Cerf, Bennett. *Try and Stop Me*. Simon and Schuster, New York. 1945.

Chaisson, Eric. *Cosmic Dawn*. Little Brown, Boston. 1981.

Chase, Stuart. 'New Energy for a New Age' in *Saturday Review*. January 22, 1955.

Chaucer, Geoffrey. *The Hous of Fame*. Clarendon Press, Oxford. 1897.

Chaucer, Geoffrey. *Troilus and Cressida* in *Great Books of the Western World*. Translated by George Philip Krappt. Volume 22. Encyclopædia Britannica, Inc., Chicago. 1952.

Chaudhuri, Haridas. *The Philosophy of Integralism*. Sri Aurobindo Pathaminder, Calcutta. 1954.

Chekhov, Anton. *Note-Book of Anton Chekhov*. Translated by S.S. Koteliansku and Leonard Woolf. B.W. Huebsch, Inc., New York. 1922.

Chesterton, G. K. *The Wisdom of Father Brown*. Dodd Mead, New York. 1930.

Chesterton, G.K. *Orthodoxy*. John Lane, London. 1909.

Chu, Steven. 'Lasers Slow Atom for Scrutiny' in *NY Times*. 13 July 1986.

Cicero. *De Natura Deorum*. Harvard University Press, Cambridge. 1955.

Clark, Ronald W. *Einstein: The Life and Times*. World Publishers, New York. 1971.

Clarke, Arthur C. *The Lost Worlds of 2001*. Gregg Press, Boston. 1972.

Clarke, Arthur C. *The Nine Billion Names of God*. Harcourt, Brace & World, Inc., New York. 1967.

Clemence, G.M. 'Time and its Measurement' in *The American Scientist*. Volume 40, Number 2. April 1952.

Cleugh, Mary F. *Time and Its Importance in Modern Thought*. Methuen & Co., Ltd., London. 1937.

Coats, R.H. 'Science and Society' in *Journal of the American Statistical Society*. Volume 34, Number 205. March 1939.

Cohen, Bernard I. *The Birth of A New Physics*. Norton, New York. 1885.

Cohen, Morris. *A Preface to Logic*. H. Holt and Company, New York. 1944.

Cole, A.D. 'Recent Evidence for the Existence of the Nucleus Atom' in *Science*. Volume 41. January 15, 1915.

Cole, K.C. 'On Imagining the Unseeable' in *Discovery*. Volume 3, Number 12. December 1982.

Coleridge, Samuel Taylor. *The Complete Poetical Works of Samuel Taylor Coleridge*. Volume I. At the Clarendon Press, Oxford. 1912.

Coles, Robert. *The Old Ones of New Mexico*. University of New Mexico Press, Albuquerque. 1973.

Collingwood, R.G. *Speculum Mentis*. At the Clarendon Press, Oxford. 1924.

Colodny, Robert G. *From Quarks to Quasars: philosophical problems of modern physics*. University of Pittsburgh Press, Pittsburgh. 1986.

Colton, Charles C. *Lacon*. William Gowans, New York. 1849.

Comte, Augustus. *The Positive Philosophy*. John Chapman, London. 1853.

Conger, George Perrigo. *A World of Epitomizations*. Princeton University Press, Princeton. 1931.

Conveney, Peter and Highfield, Roger. *The Arrow of Time*. Fawcett Columbine, New York. 1990.

Copernicus, Nicolaus. *On the Revolutions of the Heavenly Spheres* in *Great Books of the Western World*. Translated by Charles Glenn Wallis. Volume 16. Encyclopædia Britannica, Inc., Chicago. 1952.

Cort, David. *Social Astonishments*. The Macmillan Company, New York. 1963.

Courant, R. and Hilbert, D. *Methods of Mathematical Physics*. Volume I. Interscience, New York. 1953.

Cowper, William. *The Complete Poetical Works of William Cowper*. Oxford University Press, London. 1913.

Crane, Hart. *The Collected Poems of Hart Crane*. Liverlight Publishing Corporation, New York. 1946.

Crane, Stephen. *The Collected Poems of Stephen Crane*. Alfred A. Knopf, New York. 1922.

Crawford, Franzo H. *Introduction to the Science of Physics*. Harcourt, Brace & World, Inc., New York. 1968.

Crease, Robert and Mann, Charles C. *The Second Creation: Makers of the Revolution in Twentieth-Century Physics*. The Macmillan Company, New York. 1986.

Crew, Henry. *General Physics*. The Macmillan Company, New York. 1927.

Crowley, Aleister. *The Confessions of Aleister Crowley*. Hill and Wang, New York. 1969.

d'Abro, A. *The Evolution of Scientific Thought from Newton to Einstein*. Dover Publications, Inc., New York. 1950.

D'Avenant, Sir William. *Salmacida Spolia*. Printed by T.H. for Thomas Walkley, London. 1639.

D'Israeli, Isaac. *Literary Character of Men of Genius*. F. Warne, London. 1881.

da Vinci, Leonardo. *The Notebooks of Leonardo da Vinci*. Volume 1. Dover Publications, New York. 1970.

Dalton, John. *A New System of Chemical Philosophy*. Philosophical Library, New York. 1964.

Dante. *The Divine Comedy of Dante Alighieri* in *Great Books of the Western World*. Translated by Charles Eliot Norton. Volume 21. Encyclopædia Britannica, Inc., Chicago. 1952.

Dantzig, Tobias. *Number: The Language of Science*. The Macmillan Company, New York. 1954.

Darrow, Karl K. 'Some Contemporary Advances in Physics V' in *Bell Systems Technical Journal*. Volume 3. 1924.

Darwin, Francis. *The Life and Letters of Charles Darwin*. Volumes I and II. D. Appleton and Company, New York. 1896.

Dash, J. *A Life of One's Own; Three Women and the Men They Married*. Harper & Row, New York. 1973.

Davies, J.T. *The Scientific Approach*. Academic Press, London. 1965.

Davies, Paul. *About Time*. Simon & Schuster. New York. 1995.

Davies, Paul. *God and the New Physics*. Simon and Schuster, New York. 1983.

Davies, Paul. *Other Worlds*. Simon and Schuster, New York. 1980.

Davies, Paul. *Superforce*. Simon and Schuster, New York. 1984.

Davies, Paul. *The Edge of Infinity*. Simon and Schuster, New York. 1981.

Davies, Paul. *The Last Three Minutes*. Basic Books, New York. 1994.

Davies, P.C.W. and Brown, Julian. *Superstrings: A Theory of Everything?* Cambridge University Press, Cambridge. 1988.

Day, Roger E. 'Fantasy of Glass' in *The Physics Teacher*. Volume 3, Number 6. September 1965.

de Finad, J. *A Thousand Flashes of French Wit, Wisdom, and Wickedness*. D. Appleton and Company, New York. 1880.

de Fontenelle, M. *Conversations on the Plurality of Worlds*. J. Cundee, London. 1803

de Morgan, Augustus. *A Budget of Paradoxes*. Volumes I and II. The Open Court Publishing Co., Chicago. 1915.

de Saint-Exupéry, Antoine. *The Little Prince* translated by Katherine Woods. Harcourt Brace Jovanovich, Publishers, New York. 1971.

Dee, John. *John Dee on Astronomy*. Translated by Wayne Shumaker. University of California Press, Berkeley. 1978.

Deming, William Edward. *Some Theory of Sampling*. John Wiley & Sons, Inc., New York. 1960.

Deming, William Edwards. 'On the Classification of Statistics' in *The American Statistician*. Volume 2, Number 2. April 1948.

Descartes, René. *Principles of Philosophy*. Translated by Blair Reynolds. E. Mellen Press, Lewiston. 1988.

Dewey, John. *Reconstruction in Philosophy*. H. Holt & Company, New York. 1920.

DeWitt, Cecile. *Relativity, Groups and Topology*. Gordon and Breach, Science Publishers, New York. 1964.

DeWitt, Cecile M. and Wheeler, John A. *Battelle Rencontres*. W.A. Benjamin, Inc., New York. 1968

Dick, Thomas. *The Works of Thomas Dick, LL.D.* Volumes VII, IX and X. R. Worthington, New York. 1884.

Dickens, Charles. *Pickwick Papers*. Chapman & Hall, London. 1836.

Dickinson, Emily. *Poems of Emily Dickinson*. Heritage Press, New York. 1952.

Dickson, Frank. *Quote*. Volume 52, Number 3. July 17, 1966.

Dickson, Paul. *The Official Rules*. Delacorte Press, New York. 1978.

DiCurio, Robert. 'Physics inspires the Muses' in *The Physics Teacher*. December 1978.

Dilorenzo, Kirk. 'What's Physics?' in *The Physics Teacher*. Volume 14, Number 5. May 1976.

Dirac, Paul Adrien Maurice. 'Quantisized Singularities in the Electromagnetic Field' in *Proceedings of the Royal Society*. Series A, Volume 133. 1931.

Dirac, P.A.M. *The Principles of Quantum Mechanics*. Clarendon Press, Oxford. 1958.

Disraeli, Benjamin. *Tancred*. Peter Davies, London. 1927.

Donne, John. *The Complete English Poems of John Donne*. Everyman's Library, London. 1985.

Dostoevsky, Fyodor Mikhailovich. *The Brothers Karamozov*. The Macmillan Company, New York. 1955.

Doyle, Arthur Conan. *The Complete Sherlock Holmes*. Garden City Publishing Co., New York. 1930.

Dryden, John. *The Poetical Works of Dryden*. Macmillan, New York. 1904.

Du Noüy, Pierre L. *The Road to Reason*. Longman & Green, Toronto. 1949.

Duhem, Pierre. *The Aim and Structure of Physical Theory*. Atheneum, New York. 1977.

Duhem, P. 'Quelques réflexions au sujet de la physique expérimentale' in *Revue des Questions scientifiques*. Volume 36. 1897.

Duhem, P. *Théorie physique; son objet—son structure*. Paris. 1906.

Dukas, Helen and Hoffmann, Banesh. *Albert Einstein, The Human Side: New Glimpses from His Archives*. Princeton University Press, Princeton. 1979.

Dunne, Finley Peter. *Mr. Dooley At His Best*. Charles Scribner's Sons, New York. 1938.

Durack, J.J. 'Ions Mine' in *The American Physics Teacher*. June 1939.

Durant, Will. *The Story of Philosophy*. Garden City Publishing Co., Garden City. 1933.

Durgin, Richard. *Conversations with Jorge Luis Borges*. Holt, Rinehart and Winston, New York. 1969.

Dürrenmatt, Friedrich. *The Physicists*. Translated by James Kirkup. Grove Press, Inc., New York. 1964.

Dylan, Bob. 'The Two Lives of Bob Dylan' in *Newsweek*. 8 December. 1985.

Dyson, Freeman. *Infinite in all Directions*. Harper and Row, Publishers, New York. 1988.

Dyson, Freeman. *Some Strangeness in the Proportion*. Edited by H. Ward. Addison Wesley, Reading. 1980.

Dyson, Freeman. 'Energy in the Universe' in *Scientific American*. Volume 224, Number 3. 1971.

Eco, Umberto. *Foucault's Pendulum*. Harcourt Brace Jonanovich, San Diego. 1989.

Eddington, Sir Arthur Stanley. *New Pathways in Science*. University of Michigan Press, Ann Arbor. 1959.

Eddington, Sir Arthur Stanley. *Space, Time and Gravitation*. Harper & Row, Publishers, New York. 1959.

Eddington, Sir Arthur Stanley. *Stars and Atoms*. Yale University Press, New Haven. 1927.

Eddington, Sir Arthur Stanley. *The Expanding Universe*. The University Press, Cambridge. 1933.

Eddington, Sir Arthur Stanley. *The Internal Constitution of Stars*. At the University Press, Cambridge. 1930.

Eddington, Sir Arthur Stanley. *The Mathematical Theory of Relativity*. At the University Press, Cambridge. 1954.

Eddington, Sir Arthur Stanley. *The Nature of the Physical World*. Cambridge University Press. 1928.

Eddington, Sir Arthur Stanley. *The Philosophy of Physical Science*. Cambridge University Press, London. 1949.

Edgeworth, Francis Ysidro. 'On the use of the Theory of Probabilities in Statistics Relating to Society' in *Journal of the Royal Statistical Society*. January 1913.

Editor, The Louisville Journal. *Prenticeana*. Derby & Jackson, New York. 1860.

Edmonds, Charles. *Poetry of the Anti-Jacobian*. Sampson Low, Marstow, Searle, & Rivington, London. 1890.

Einstein, Albert. 'Atomic Education Urged by Einstein' in *NY Times*. 25 May 1946.

Einstein, Albert. *Autobiographical Notes*. Translated by Paul Arthur Schlipp. Open Court, La Salle. 1979.

Einstein, Albert. *Ideas and Opinions*. Bonaza Books, New York. 1954.

Einstein, Albert. *Out of My Later Years*. Thames and Hudson, London. 1950.

Einstein, Albert. *Relativity, The Special and General Theory*. Translated by Robert W. Lawson. Henry Holt and Co., New York. 1920.

Einstein, Albert. *Sidelights on Reality*. E.P. Dutton, New York. 1922.

Einstein, Albert. *The Evolution of Physics*. Simon and Schuster, New York. 1938.

Einstein, Albert. *The Principle of Relativity*. University of Calcutta, Calcutta. 1920.

Einstein, Albert. *The World As I See It*. Philosophical Library, New York. 1949.

Eliot, George. *Theophrastus Such*. Crowell, New York. 189?.

Eliot, T.S. *Burnt Norton*. Faber & Faber, London. 1941.

Eliot, T.S. *The Complete Poems and Plays, 1909–1950*. Harcourt, Brace and World, New York. 1962.

Emerson, Ralph Waldo. *The Works of Ralph Waldo Emerson*. Houghton Mifflin Company, Boston. 1903.

Esar, Evan. *Esar's Comic Dictionary*. Doubleday and Company, Inc., Garden City. 1983.

Evans, Bergen. *The Natural History of Nonsense*. Alfred A. Knopf, New York. 1946.

Eves, Howard W. *Mathematical Circles Squared*. Prindle, Weber & Schmidt, Inc., Boston. 1972.

Fabing, Harold and Marr, Ray. *Fischerisms*. C.C. Thomas, Springfield. 1937.

Fadiman, Clifton. *The Mathematical Magpie*. Simon and Schuster, New York. 1962.

Faraday, Michael. *Experimental Researches in Electricity*. Volume I. Dover Publications, Inc., New York. 1965

Felix, Lucienne. *The Modern Aspect of Mathematics*. Translated by Julius H. Hlavaty and Fancille H. Hlavaty. Basic Books, Inc., New York. 1960.

Ferris, Timothy. *The Red Limit*. William Morrow & Co., Inc., New York. 1977.

Feynman, Richard P. *QED*. Princeton University Press, Princeton. 1985.

Feynman, Richard. *The Character of Physical Law*. British Broadcasting Corporation, London. 1965.

Feynman, Richard P., Leighton, Robert B. and Sands, Matthew. *The Feynman Lectures on Physics*. Volume 1. Addison-Wesley Publishing Company, Reading. 1964.

Feynman, Richard P. *The Feynman Lectures on Physics*. Volume 2. Addison-Wesley Publishing Company, Reading. 1964.

Feynman, Richard P. *The Feynman Lectures on Physics*. Volume 3. Addison-Wesley Publishing Company, Reading. 1964.

Field, Edward. *New and Selected Poems from the Book of My Life*. The Sheep Meadow Press, New York. 1987.

Fine, Arthur. *The Shakey Game*. The University of Chicago Press, Chicago. 1986.

Flammarion, Camille. *Popular Astronomy: A General Description of the Heavens*. Chatto & Windus, Piccadilly. 1894.

Flanders, M. and Swann, D. 'The First and Second Law' sung on their record: *At the Drop of Another Hat*. EMI Records, 1964.

Flaubert, Gustave. *Early Writings*. Translated by Robert Griffin. University of Nebraska Press, Lincoln. 1991.

Footner, Hulbert. *A Backwoods Princess*. George H. Duran Co., New York. 1926.

Ford, John. *The Broken Heart*. University of Nebraska Press, Lincoln. 1968.

Ford, Joseph. 'How Random is a Coin Toss?' in *Physics Today*. April 1983.

Fort, Charles. *The Books of Charles Fort*. Henry Holt and Company, New York. 1941.

Foster, G.C. 'Mathematical and Physical' in *Nature*. Volume 16, Number 407. August 16, 1887.

Fourier, Jean Baptiste Joseph. *Analytical Theory of Heat* in *Great Books of the Western World*. Volume 45. Encyclopædia Britannica, Inc., Chicago. 1952.

Fournier d'Albe, E.E. *The Electron Theory*. Longmans, Green and Co., London. 1906.

France, Anatole. *Anatole France Himself*. J.B. Lippincott Company, Philadelphia. 1925.

Frank, Phillip. *Modern Science and its Philosophy*. Harvard University Press, Cambridge. 1950.

Freeman, Ira M. 'Nuclear Situation Unclear' in *The Physics Teacher*. Volume 13, Number 5. May 1975.

French, A.P. *Einstein: A Centenary Volume*. Harvard University Press, Cambridge. 1979.

French, Anthony and Taylor, Edwin. *An Introduction to Quantum Physics*. Norton, New York. 1978.

Freud, Sigmund. *On Narcissism* in *Great Books of the Western World*. Translated by Cecil M. Baines. Volume 54. Encyclopædia Britannica, Inc., Chicago. 1952.

Friedman, Herbert. *The Astronomer's Universe*. W.W. Norton & Company, New York. 1990.

Fritzsch, Harald. *The Creation of Matter*. Basic Books, Inc., New York. 1984.

Frost, Robert. *The Poetry of Robert Frost*. Holt, Rinehart, and Winston, New York. 1965.

Froude, James Anthony. *Short Studies on Great Subjects*. E.P. Dutton, New York. 1930–31.

Fry, Christopher. *The Lady's Not for Burning*. Oxford University Press, New York. 1949.

Fuchs, Walter R. *Mathematics for the Modern Mind*. Macmillan Publishing, New York. 1967.

Fuller, Thomas. *The Holy and Profane States*. Little Brown, Boston. 1864.

Furth, Harold P. 'Perils of Modern Living' in *The New Yorker*. November 10, 1956.

Galileo. *Dialogues Concerning the Two Chief World Systems*. Translated by Stillman Drake. University of California Press, Berkeley and Los Angeles. 1962.

Galileo. *Dialogues Concerning Two New Sciences* in *Great Books of the Western World*. Translated by Henry Crew and Alfonso de Salvio. Volume 28. Encyclopædia Britannica, Inc., Chicago. 1952.

Gamow, George. *A Star Called the Sun*. Viking Press, New York. 1964.

Gamow, George. *Biography of Physics*. Harper & Row Publishers, New York. 1961.

Gamow, George. *Matter, Earth and Sky*. Prentice-Hall, Englewood Cliffs. 1965.

Gamow, George. *Mr. Tompkins in Paperback*. At the University Press, Cambridge. 1965.

Gamow, George. *My World Line*. The Viking Press, New York. 1970.

Gamow, George. *The Birth and Death of the Sun*. The Viking Press, New York. 1943.

Gamow, George. *Thirty Years that Shook Physics*. Doubleday & Company, Inc., Garden City. 1966.

Gamow, George. 'The Reality of Neutrinos' in *Physics Today*. July 1948.

Gamow, George. 'Any Physics Tomorrow?' in *Physics Today*. January 1949.

Gaposchkin, Cecilia Helena Payne. *Introduction to Astronomy*. Prentice-Hall, Englewood Cliffs. 1970.

Gardner, Earl Stanley. *The Case of the Buried Clock*. Grosset & Dunlap, New York. 1943.

Gardner, Earl Stanley. *The Case of the Perjured Parrot*. The Blakiston Company, Philadelphia. 1939.

Gardner, Martin. *Fads and Fallacies*. Dover Publications, Inc., New York. 1957.

Gardner, Martin. *The Ambidextrous Universe*. Basic Books, Inc., New York. 1964.

Gardner, Martin. *Time Travel and Other Mathematical Bewilderments*. W.H. Freeman and Company, New York. 1988.

Gilbert, William. *On the Loadstone and Magnetic Bodies* in *Great Books of the Western World*. Volume 28. Encyclopædia Britannica, Inc., Chicago. 1952.

Gilbert, W.S. and Sullivan, Arthur. *Patience*. Art Interchange Publication, Co., New York. 1882.

Gilman, Greer. 'Ergo' in *The Physics Teacher*. Volume 8, Number 9. December 1970.

Glasgow, Ellen. *Letters of Ellen Glasgow*. Harcourt Brace and Co., New York. 1958.

Gleick, James. *Chaos*. Viking Press, New York. 1987.

Gleick, James. *Genius: The Life and Science of Richard Feynman*. Pantheon Books, New York. 1992.

Goethe, Johann Wolfgang von. *Faust*. J. Cape & H. Smith, New York. 1937.

Goethe, Johann Wolfgang von. *Maxims and Reflections*. Macmillan and Company, London. 1893.

Graham, L.A. *Ingenious Mathematical Problems and Methods*. Dover Publications, New York. 1959.

Graves, Robert. *Life of Sir William Rowan Hamilton*. Hodges, Figgs & Co., Dublin. 1882–1889.

Greenberg, D.S. *The Politics of Pure Science*. New American Library, New York. 1967.

Greenstein, Jesse L. 'Great American Scientists: The Astronomer' in *Fortune*. Volume 61, Number 5. May 1960.

Gribbin, John. *The Omega Point: The Search for the Missing Mass and the Ultimate Fate of the Universe*. Bantam Books, New York. 1988.

Gross, David. 'On the Calculation of the Fine-Structure Constant' in *Physics Today*. December 1989.

Gudder, Stanley. *A Mathematical Journey*. McGraw-Hill Book Company, New York. 1976.

Haldane, J.B.S. *Possible Worlds and Other Papers*. Harper & Brothers Publishers, New York. 1928.

Hamilton, Edith. *The Roman Way*. W.W. Norton, New York. 1932.

Hammond, A.L. and Metz, W.D. 'Solar Energy Research: Making Solar After the Nuclear Model?' in *Science*. Volume 197. 1977.

Hanson, Norwood Russell. *Patterns of Discovery*. At the University Press, Cambridge. 1958.

Hardy, G.H. *A Mathematician's Apology*. At the University Press, Cambridge. 1967.

Hardy, Thomas. *Far from the Madding Crowd*. Macmillan, London. 1974.

Hardy, Thomas. *The Collected Poems of Thomas Hardy*. The Macmillan Company, New York. 1974.

Hardy, Thomas. *Thomas Hardy's Chosen Poems*. F. Ungar Publications, New York. 1978.

Hardy, Thomas. *Two on a Tower*. Macmillan, London. 1976.

Harrison, Edward. *Cosmology*. Cambridge University Press, London. 1961.

Harrison, Edward. *Masks of the Universe*. The Macmillan Company, New York. 1985.

Hawking, Stephen. *A Brief History of Time*. Bantam Books, New York. 1988.

Hawking, Stephen. 'The Quantum Mechanics of Black Holes' in *Scientific American*. Volume 236, Number 1. January 1977.

Haynes, Renée. 'Signs of Secrecy' in *Times Literary Supplement*. June 18, 1981.

Heaviside, Oliver. *Electromagnetic Theory*. Dover Publications, Inc., New York. 1950.

Hein, Piet. *Grooks II*. Doubleday, Garden City. 1968.

Hein, Robert. *Quest of the Singing Tree*. Henry Harrison, New York. 1938.

Heinlein, Robert A. *Assignment in Eternity*. Fantasy Press, Reading. 1953.

Heinlein, Robert A. *Orphans of the Sky*. New American Library, New York. 1963.

Heinlein, Robert A. *The Cat Who Walks Through Walls*. Putnam, New York. 1985.

Heinlein, Robert A. *Time Enough for Love*. G.P. Putnam's Sons, New York. 1973.

Heinlein, Robert A. *Time for the Stars*. Scribner, New York. 1956.

Heisenberg, Werner. *Across the Frontiers*. Harper and Row, New York. 1974.

Heisenberg, Werner. *On Modern Physics*. C.N. Potter, New York. 1961.

Heisenberg, Werner. *Philosophic Problems of Nuclear Science*. Translated by F.C. Hayes. Faber and Faber, Ltd., London. 1952.

Heisenberg, Werner. *Physics and Beyond: Encounters and Conversations*. Harper and Row, New York. 1971.

Heisenberg, Werner. *Physics and Philosophy*. Harper & Row, New York. 1958.

Heisenberg, Werner. *The Physical Principles of the Quantum Theory*. Translated by Carl Ekhart and Frank C. Hoyt. The University of Chicago Press. 1930.

Hembree, Lawrence. *Quote*. Volume 54, Number 6. April 6, 1967.

Henry, M.L. 'The Lunar Landing and the US–Soviet Equation' in *Bulletin of the Atomic Scientists*. Volume 25, Number 7. September 1969.

Herbert, George. *The Works of George Herbert*. Thomas Y. Crowall & Co., New York. No date.

Herbert, Nick. *Quantum Reality*. Anchor Press, Garden City. 1985.

Herbert, Nick. *Faster than Light*. New American Library, New York. 1988.

Herschel, Sir John. *Herschel at the Cape*. Edited by David S. Evans, Terence J. Deeming, Betty Hall Evans, and Stephen Goldfarb. University of Texas Press, Austin. 1969.

Hertz, H. *Electric Waves; Being Researches on the Propagation of Electric Action with Finite Velocity through Space*. Macmillan & Co., London. 1893.

Hill, Thomas. 'The Imagination in Mathematics' in *North American Review*. Volume 85, Number 176. July 1857.

Hodgson, Ralph. *Collected Poems*. The Macmillan Company, Ltd., London. 1961.

Hoffmann, Banesh. *The Strange Story of the Quantum*. Dover Publications, Inc., New York. 1959.

Holman, Mrs. Jesse B. *The Zodiac, The Constellations and the Heavens*. Printed by E.L. Steck, Co., Austin. 1924.

Holmes, John Haynes. *A Sensible Man's View of Religion*. Harper & Brothers Publishers, New York. 1932.

Holmes, O.W. *The Autocrat of the Breakfast Table*. Houghton Mifflin, Boston. 1894.

Holmes, O.W. *The Poet at the Breakfast Table*. Houghton Mifflin, Boston. 1892.

Holton, G. *Thematic Origins of Scientific Thought*. Harvard University Press, Cambridge. 1973.

Homer. *The Iliad* in *Great Books of the Western World*. Volume 4. Encyclopædia Britannica, Inc., Chicago. 1952.

Homer. *The Odyssey* in *Great Books of the Western World*. Volume 4. Encyclopædia Britannica, Inc., Chicago. 1952.

Hooke, Robert. *Micrographia*. Weinheim, Cramer, New York. 1961.

Hopkins, Gerard Manley. *The Poetical Works of Gerard Manley Hopkins*. Clarendon Press, Oxford. 1990.

Horton, F. 'The Radium Atom' in *The American Physics Teacher*. Volume 7, Number 3. June 1939.

Howard, Neale E. *The Telescope Handbook and Star Atlas*. Crowell, New York. 1975.

Hoyle, Fred. *Frontiers of Astronomy*. Harper, New York. 1955.

Hoyle, Fred. *Galaxies, Nuclei and Quasars*. Heinemann, London. 1965.

Hoyle, Fred. *October the First is Too Late*. Harper and Row, New York. 1966.

Hoyle, Fred. *The Black Cloud*. Harper, New York. 1957.

Hoyle, Fred. *The Nature of the Universe*. Harper & Brothers, New York. 1950.

Hoyle, Fred and Hoyle, Geoffrey. *Into Deepest Space*. Harper & Row, New York. 1974.

Hoyle, Fred and Hoyle, Geoffrey. *The Inferno*. Harper & Row, New York. 1973.

Hubble, Edwin. *The Nature of Science and other Lectures*. Huntington Library, San Marino. 1954.

Hubble, Edwin. *The Realm of the Nebulae*. Yale University Press, New Haven. 1982.

Huebner, Jay S. 'What's Physics' in *The Physics Teacher*. Volume 14, Number 5. May 1976.

Huggins, Sir William. *The Scientific Papers of Sir William Huggins*. W. Wesley and Son, London. 1909.

Huntington, Edward V. *The Continuum*. Harvard University Press. 1917.

Hutten, Ernest H. *The Language of Modern Physics*. George Allen and Unwin, Ltd., London. 1956.

Huxley, Aldous. *Literature and Science*. Harper and Row, New York. 1963.

Huxley, Aldous. *Point Counter Point*. Harper & Brothers Publishers, New York. 1928.

Huxley, Julian. *Essays in Popular Science*. Alfred A. Knopf, New York. 1927.

Huxley, Thomas. *Man's Place in Nature*. University of Michigan Press, Ann Arbor. 1959.

Huxley, Thomas. *The Life and Letters of Thomas Huxley*. D. Appleton, New York. 1901.

Huxley, Thomas. *Lay Sermons, Addresses, and Reviews*. D. Appleton & Company, New York. 1871.

Huxley, T.H. *Collected Essays (Darwina)*. Volume II. D. Appleton and Company, New York. 1903.

Huygens, Christiaan. *The Celestial Worlds Discover'd*. Frank Cass & Co., Ltd., London. 1968.

Huygens, Christiaan. *Treatise on Light* in *Great Books of the Western World*. Volume 34. Encyclopædia Britannica, Inc., Chicago. 1952.

Iannelli, Richard. *The Devil's New Dictionary*. Citadel Press, Secaucus. 1983.

Infeld, Leopold. *Quest—An Autobiography*. Chelsea Publishing Company, New York. 1980.

Jaki, S.L. 'Goethe and the Physicists' in *American Journal of Physics*. Volume 37. July–December 1969.

Jeans, Sir James H. *Astronomy and Cosmogony*. Dover Publications, Inc., New York. 1961.

Jeans, Sir James. *Physics and Philosophy*. University of Michigan Press, Ann Arbor. 1958.

Jeans, Sir James. *The Mysterious Universe*. The Macmillan Company, New York. 1948.

Jeans, Sir James. *The Universe Around Us*. The Macmillan Company, New York. 1929.

Jeavons, W.S. *The Principles of Science*. Dover Publications, New York. 1958.

Jeffers, Robinson. *The Beginning and the End*. Random House, New York. 1954.

Jefferson, Thomas. *Notes on the State of Virginia*. Norton, New York. 1954.

Jefferys, William H. and Robbins, R. Robert. *Discovering Astronomy*. John Wiley & Sons, New York. 1981.

Jeffreys, H. and Jeffreys, B.S. *Methods of Mathematical Physics*. University Press, Cambridge. 1956.

Jennings, Elizabeth. *Selected Poems*. Carcanet, Manchester. 1979.

Jespersen, James and Fitz-Randolph, Jane. *From Quarks to Quasars*. University of Pittsburgh Press. 1986.

Jones, R.V. 'The Natural Philosophy of Flying Saucers' in *Physics Bulletin*. Volume 19, July 1968.

Joubert, Joseph. *Pensées*. Brentano's, New York. 1928.

Kac, Mark. *Statistical Independence in Probability Analysis and Number Theory*. Wiley, New York. 1959.

Kaku, Michio. *Hyperspace*. Oxford University Press, New York. 1994.

Kant, Immanuel. *Kant's Cosmogony*. Translated by W. Hastie. James Maclehose and Sons, Glasgow. 1900.

Kant, Immanuel. *The Critique of Judgment* in *Great Books of the Western World*. Translated by James Creed Meredoth. Volume 42. Encyclopædia

Britannica, Inc., Chicago. 1952.

Kant, Immanuel. *The Critique of Pure Reason* in *Great Books of the Western World*. Translated by J.M.D. Meiklejohn. Volume 42. Encyclopædia Britannica, Inc., Chicago. 1952.

Kantha, Sachi Sri. *An Einstein Dictionary*. Greenwood Press, Westport. 1996.

Kasner, Edward and Newman, James R. *Mathematics and the Imagination*. Simon and Schuster, New York. 1967.

Kelvin, William Thomson, Baron. *Baltimore Lectures on Molecular Dynamics, and the Wave Theory of Light*. Cambridge University Press, London. 1904.

Kelvin, William Thomson, Baron. *Popular Lectures and Addresses*. Macmillan and Co., London. 1891–94.

Keyser, C.J. *The Human Worth of Rigorous Thinking*. Columbia University Press, New York. 1916.

King, Alexander. *I Should have Kissed Her More*. Simon and Schuster, New York. 1961.

Kippenhahn, Rudolf. *Light from the Depths of Time*. Springer-Verlag, New York. 1987.

Kirkup, James. *Omens of Disaster*. University of Salzburg, Salzburg. 1996.

Klein, H. Arthur. *The World of Measurements*. Simon and Schuster, New York. 1974.

Kline, Morris. *Mathematics and the Physical World*. Thomas Y. Crowell, Company. New York. 1959.

Kline, Morris. *Mathematical Thought from Ancient to Modern Times*. Oxford University Press, New York. 1972.

Kline, Morris. *Mathematics in Western Culture*. Oxford University Press, New York. 1953.

Koestler, Arthur. *The Act of Creation*. Pan Books, London. 1964.

Koestler, Arthur. *The Sleepwalkers*. The Macmillan Company, New York. 1968.

Koyre, Alexander. *From the Closed World to the Infinite Universe*. Johns Hopkins Press, Baltimore. 1957.

Kraus, Karl. *Half-Truths & One-and-a-Half Truths*. Engenda Press, Montreal. 1976.

Krauss, Lawrence M. *Fear of Physics*. Basic Books, New York. 1993.

Krutch, Joseph Wood. *The Twelve Seasons*. William Sloane Associates, Publishers, New York. 1949.

Kuh, Katharine. *The Artist's Voice*. Harper & Row, Publishers, New York. 1962.

Kuhn, Thomas. *The Structure of Scientific Revolutions*. University of Chicago Press, Chicago. 1970.

Kunitz, Stanley. *Selected Poems 1928–1958*. Little, Brown and Company, Boston. 1958.

Laplace, Pierre Simon. *Celestial Mechanics*. Chelsea Publishing Co., Bronx. 1966.

Larrabee, Eric. 'Easy Road to Culture, Sort of' in *Humor From Harper's*. Harper & Brothers, New York. 1961.

Latham, Peter M. *The Collected Works of Dr. P.M. Latham*. The New Sydenham Society, London. 1876–1878.

Launer, Robert L. and Wilkinson, Graham N. *Robustness in Statistics: Proceedings of a Workshop*. Academic Press, New York. 1979.

Law, Frederick Houk. *Science in Literature*. Harper & Brothers Publishers, New York. 1929.

Leacock, Stephen. *Literary Lapses*. Gazette Printing Company, Limited, Montreal. 1910.

Leacock, Stephen. *Winnowed Wisdom: A New Book of Humor*. Dodd, Mead & Company, New York. 1926.

Lederman, Leon. *The God Particle*. Houghton Mifflin Company, Boston. 1993.

Lee, A. 'Gravity' in *The Physics Teacher*. Volume 22, Number 7. October 1984.

Lehman, Robert C. 'Eureka' in *The Physics Teacher*. Volume 21, Number 2. February 1983.

Lemaître, G. 'The Beginning of the World from the Point of View of Quantum Theory' in *Nature*. Volume 127. May 9, 1931.

Lemaître, G. *The Primeval Atom*. Van Nostrand, New York. 1951.

Lerner, Eric J. *The Big Bang Never Happened*. Time Books, New York. 1991.

Levi, Primo. *Primo Levi Collected Poems*. Faber & Faber, London. 1988.

Levy, David H. *The Man Who Sold the Milky Way*. The University of Arizona Press, Tucson. 1993.

Levy, Hyman. *The Universe of Science*. The Century Company, New York. 1933.

Lewis, C.S. *The Pilgrim's Regress: An Allegorical Apology for Christianity, Reason and Romanticism*. Beerdmans, Grand Rapids. 1943.

Lewis, Gilbert N. *The Anatomy of Science*. Yale University Press, New Haven. 1926.

Lichtenberg, Georg C. *Lichtenberg: Aphorisms & Letters*. Translated by Franz Mautner and Henry Hatfield. Jonathan Cape, London. 1969.

Liebson, Morris. 'Physics inspires the Muses' in *The Physics Teacher*. December 1978.

Lillich, Robert. 'My Fair Physicist' in *The Physics Teacher*. December 1968.

Lindsay, R.B. 'The Broad Point of View in Physics' in *The Scientific Monthly*. February 1932.

Locke, John. *An Essay Concerning Human Understanding*. At the Clarendon Press, Oxford. 1956.

Lodge, Oliver. *Ether and Reality*. George H. Doran Company, New York. 1925.

Lodge, O. 'The Ether and Electrons' in *Supplement to Nature*. Volume 112, Number 2805. August 4, 1923.

Longair, Malcolm. *Alice and the Space Telescope*. The Johns Hopkins University Press, Baltimore. 1989.

Longfellow, Henry Wadsworth. *The Poetical Works of Henry Wadsworth Longfellow*. Ticknor and Fields, Boston. 1867.

Lonsdale, Dame Kathleen. *Crystals and X-Rays*. D. Van Nostrand Company, Inc., New York. 1949.

Lorentz, H.A., Einstein, A., Minkowski, H. and Weyl, H. *The Principle of Relativity*. Dover Publications, Inc. 1923.

Lorenz, Konrad. *On Aggression*. Harcourt, Brace & World, New York. 1966.

Lubbock, Constance A. *The Herschel Chronicle*. The Macmillan Company, New York. 1933.

Lucan. *Pharsalia*. Penguin Books Ltd., Harmondsworth. 1956.

Lucretius. *On the Nature of Things* in *Great Books of the Western World*. Translated by H.A.J. Munro. Volume 12. Encyclopædia Britannica, Inc., Chicago. 1952.

Luminet, Jean-Pierre. *Black Holes*. University Press, Cambridge. 1987.

Lundberg, Derek. 'dB or not dB?' in *Physics World*. September 1993.

McAleer, Neil. *The Mind Boggling Universe: A Dazzling Scientific Journey Through Distant Space and Time*. Doubleday, Garden City. 1987.

MacGowan, Roger A. and Ordway, Frederick I., III. *Intelligence in the Universe*. Prentice-Hall, Englewood Cliffs. 1966.

Mach, E. *History and Root of the Principle of the Conservation of Energy*. The Open Court Publishing Company, Chicago. 1911.

Mackay, Charles. *The Poetical Works of Charles Mackay*. G. Routledge, London. 1857.

Maimondides, Moses. *The Guide of the Perplexed*. Translated by M. Friedlander. George Routledge & Sons, Ltd., London. 1919.

Mallove, Eugene F. *The Quickening Universe*. St. Martin's Press, New York. 1987.

Mamula, Karl C. 'Physics Notes' in *The Physics Teacher*. Volume 13, Number 4. April 1975.

Mann, Thomas. *The Magic Mountain*. Translated by H.T. Lowe-Porter. Alfred A. Knopf, New York. 1945.

Maor, Eli. *To Infinity and Beyond: A Cultural History of the Infinite*. Birkhäuser, Boston. 1987.

March, Robert H. *Physics for Poets*. McGraw-Hill Book Company, New York. 1978.

Marlowe, Christopher. *Tamburlaine the Great*. W. Heinemann, London. 1951.

Marquis, Don. *the lives and times of archy & mehitabel*. Doubleday Doran & Co., Inc., Garden City. 1933.

Martin, Florence Holcomb. 'The Riddle of the Skies' in *The Scientific Monthly*. August 1952.

Marton, Ladislaus. 'Alice in Electronland' in *American Scientist*. Volume 31, Number 3. July 1943.

Maxwell, James Clerk. *The Scientific Papers of James Clerk Maxwell*. Volume 2. Edited by W.D. Niven. Dover Publications, New York. 1965.

Maxwell, James Clerk. *The Theory of Heat*. Longmans, London. 1871.

Mehra, J. *The Physicist's Conception of Nature*. Reidel, Boston. 1973.

Mellor, J.W. *Higher Mathematics for Students of Chemistry and Physics*. Dover Publications, New York. 1955.

Melneckuk, Theodore. 'The Hunting of the Quark' in *The Physics Teacher*. Volume 7, Number 7. October 1969.

Mencken, H.I. *Minority Report: H.L. Mencken's Notebook*. Alfred A. Knopf, New York. 1956.

Metsler, William. 'The Cowboy's Lament' in *The Physics Teacher*. February 1977.

Michelmore, Peter. *Einstein, profile of the man*. Dodd, Mead, New York. 1962.

Miller, Henry. *Black Spring*. Grover Press, Inc., New York. 1963.

Millikan, Robert Andrews. *The Electron*. The University of Chicago Press, Chicago. 1917.

Milton, John. *Il Penseroso*. The Limited Editions Club, New York. 1954.

Milton, John. *Paradise Lost* in *Great Books of the Western World*. Volume 32. Encyclopædia Britannica, Inc., Chicago. 1952.

Miner, Virginia Scott. 'Physics inspires the Muses' in *The Physics Teacher*. Volume 16, Number 9. December 1978.

Minkowski, Hermann. *The Principle of Relativity*. The University of Calcutta, Calcutta. 1920.

Misner, Charles W., Thorne, Kip, and Wheeler, John A. *Gravitation*. W.H. Freeman, San Francisco. 1973.

Montague, W.P. 'The Einstein Theory and the Possible Alternative' in *Philosophical Review*. Volume 33, Number 2. March 1924.

Montaigne, Michel Eyquen. *The Essays of Michel de Montaigne*. Alfred A. Knopf, New York. 1934.

Moore, Ruth. *Niels Bohr: The Man, His Science, & The World They Changed*. Alfred A. Knopf, New York. 1966.

Moulton, F.R. *Introduction to Astronomy*. The Macmillan Company, New York. 1922.

Murchie, Guy. *The Seven Mysteries of Life*. Houghton Mifflin Company, Boston. 1978.

Myrdal, Gunnar. *Objectivity in Social Research*. Pantheon Books, New York. 1969.

Nabokov, Vladimir. *Bend Sinister*. Henry Holt and Company, New York. 1947.

Nabokov, Vladimir. *Ada or Ardor: A Family Chronicle*. McGraw-Hill Book Company, New York. 1969.

Narlikar, Jayant. *Violent Phenomena in the Universe*. Oxford University Press, Oxford. 1982.

Neugebauer, Otto. *The Exact Sciences in Antiquity*. Harper & Brothers, New York. 1962.

Newman, James R. *The World of Mathematics*. Simon and Schuster, New York. 1956.

Newton, Sir Isaac. *Mathematical Principles of Natural Philosophy* in *Great Books of the Western World*. Volume 34. Encyclopædia Britannica, Inc., Chicago. 1952.

Newton, Sir Isaac. *Opticks* in *Great Books of the Western World*. Volume 34. Encyclopædia Britannica, Inc., Chicago. 1952.

Nichol, John. *Views of the Architecture of the Heavens*. W. Tait, Edinburgh. 1845.

Nicholson, Norman. *Collected Poems*. Faber & Faber, London. 1994.

Nicholson, Norman. *The Pot Geranium*. Faber & Faber, London. 1954.

Nietzsche, Friedrich. *Beyond Good and Evil*. Vintage Books, New York. 1966.

Nietzsche, Friedrich. *The Gay Science*. Vintage Books, New York. 1974.

Nizer, Louis. *My Life in Court*. Random House, New York. 1964.

Noll, Ellis D. 'What's Physics?' in *The Physics Teacher*. Volume 14, Number 5. May 1976.

Nordman, Charles. *Einstein and the Universe*. Henry Holt & Co., New York. 1922.

Noyes, Alfred. *Watchers of the Sky*. Frederick A. Stokes Company, New York. 1922.

Nuegerbauer, Otto. *The Exact Sciences in Antiquity*. Brown University Press, Providence. 1957.

Nye, Bill. *Remarks*. G.P. Brown Publishing Co., Chicago. 1888.

O'Casey, Sean. *Juno and the Paycock*. Samuel French, New York. 1899.

Ohanian, Hans C. *Gravitation and Spacetime*. Norton, New York. 1976.

Olcott, William Tyler. *Star Lore of All Ages*. G.P. Putnam's Sons, New York. 1911.

Oman, John. *The Natural and the Supernatural*. The Macmillan Company, New York. 1931.

O'Neil, W.M. *Fact and Theory*. Sydney University Press, Sydney. 1969.

Oppenheimer, J. Robert. *Science and the Common Understanding*. Simon and Schuster, New York. 1954.

Oppenheimer, J. Robert. 'Expiation' in *Time*. Volume 51, Number 8. February 23, 1948.

Osserman, Robert. *Poetry of the Universe: A Mathematical Exploration of the Universe*. Anchor Books, New York. 1995.

Ovid. *Metamorphoses*. Duke University Press, Durham. 1968.

Pagels, Heinz. *The Cosmic Code*. Simon and Schuster, New York. 1982.

Pagels, Heinz. *Perfect Symmetry*. Simon and Schuster, New York. 1985.

Pais, Abraham. 'Particles' in *Physics Today*. Volume 21, May 1968.

Pais, Abraham. *Inward Bound*. Clarendon Press, Oxford. 1986.

Panda, N.C. *Maya in Physics*. Motilal Bonarsidass Publishers, Delhi. 1991.

Panofsky, Wolfgang. 'Particle Substructure: A Common Theme of Discovery in this Century' in *Contemporary Physics*. Volume 20, Number 1. 1982.

Parker, E.N. *Cosmical Magnetic Fields*. Secker & Warburg, London. 1963.

Pasachoff, Jay M. 'Pulsars in Poetry' in *Physics Today*. Volume 22, Number 2. February 1969.

Pascal, Blaise. *Pensées* in *Great Books of the Western World*. Translated by W.F. Trotter. Volume 33. Encyclopædia Britannica, Inc., Chicago. 1952.

Pascal, Blaise. *Scientific Treatises* in *Great Books of the Western World*. Translated by Richard Scofield. Volume 33. Encyclopædia Britannica, Inc., Chicago. 1952.

Pasternak, Boris. *Boris Pasternak: Fifty Poems*. Translated by Lydia Pasternak Slater. George Allen & Unwin Ltd., London. 1963.

Patten, W. *The Grand Strategy of Evolution*. R.G. Badger, Boston. 1920.

Pearson, Karl. *Life, Letters and Labours of Francis Galton*. University Press, Cambridge. No date.

Peebles, P.J.E. *Physical Cosmology*. Princeton University Press, Princeton. 1971.

Penrose, Roger. *The Emperor's New Mind*. Oxford University Press, Oxford. 1989.

Petit, Jean-Pierre. *Euclid Rules OK?* John Murray, London. 1982.

Petroski, Henry. 'The Mathematical Physicist' in *Southern Humanities Review*. Volume 8, Number 2. 1972.

Pickover, Clifford A. *Keys to Infinity*. Wiley & Sons, New York. 1995.

Pierce, C.S. 'The Architecture of Theories' in *The Monist*. Volume 11, Number 2. January 1891.

Pile, Stephen. *The Book of Heroic Failures*. Routledge & Kegan Paul, London. 1979.

Planck, Max. *A Survey of Physics*. Methuen & Co., Ltd., London. 1925.

Planck, Max. *Where is Science Going?* Translated by James Murphy. W.W. Norton, New York. 1977.

Planck, Max. *Scientific Autobiography and Other Papers*. Philosophical Library, New York. 1949.

Planck, Max. *The Universe in the Light of Modern Physics*. George Allen & Unwin Ltd., London. 1931.

Plato. *Phaedrus* in *Great Books of the Western World*. Translated by Benjamin Jowett. Volume 7. Encyclopædia Britannica, Inc., Chicago. 1952.

Plato. *The Republic* in *Great Books of the Western World*. Translated by Benjamin Jowett. Volume 7. Encyclopædia Britannica, Inc., Chicago. 1952.

Plato. *Timaeos* in *Great Books of the Western World*. Translated by Benjamin Jowett. Volume 7. Encyclopædia Britannica, Inc., Chicago. 1952.

Plotz, Helen. *Imagination's Other Place*. Crowell, New York. 1955.

Poe, Edgar Allan. *The Complete Works of Edgar Allan Poe*. Volume XVI. 'Eureka'. Fred deFau & Co., New York. 1902.

Poincaré, Henri. *The Foundations of Science*. The Science Press, New York. 1921.

Polanyi, Michael. *Personal Knowledge*. The University of Chicago Press, Chicago. 1958.

Polya, G. *How to Solve It*. Princeton University Press, Princeton. 1945.

Ponomarev, L.I. *The Quantum Dice*. IOP Publishing, Bristol. 1993.

Pope, Alexander. *The Complete Poetical Works of POPE*. Edited by Henry W. Boynton. Houghton Mifflin and Company, Boston. 1931.

Popper, Karl. *Conjectures and Refutations: The Growth of Scientific Knowledge*. Harper and Row, New York. 1965.

Popper, Karl. *Realism and the Aim of Science*. Rowman and Littlefield, Totowa. 1983.

Popper, Karl. *The Logic of Scientific Discovery*. Hutchinson, London. 1959.

Popper, Karl. *The Open Universe*. Rowman and Littlefield, Totowa. 1982.

Prior, Matthew. *The Literary Works of Matthew Prior*. Volume VI. At the Clarendon Press, Oxford. 1959.

Rainich, G.Y. 'Analytic Function and Mathematical Physics' in *Bulletin of the American Mathematical Society*. October 1931.

Raman, V.V. 'A Fable for Physicists' in *The Physics Teacher*. Volume 18, Number 7. October 1990.

Raymo, Chet. *The Soul of the Night*. Prentice-Hall, Englewood Cliffs. 1985.

Reichen, Charles-Albert. *A History of Astronomy*. Hawthorn Books Inc., New York. 1963.

Reichenbach, Hans. *The Philosophy of Space & Time*. Dover Publications, Inc., New York. 1958.

Reid, Constance. *Hilbert*. Springer-Verlag, New York. 1970.

Renya, Ruth. *The Philosophy of Matter in the Atomic Era*. Asia Publishing House, Bombay. 1962.

Rilke, Rainer Maria. *The Duino Elegies*. Translated by Leslie Norris and Alan Keele. Camden House, Inc., Columbia. 1993.

Rilke, Rainer Maria. *Sonnets to Orpheus*. Translated by Stephen Mitchell. Simon and Schuster, New York. 1985.

Rindler, Wolfgang. *Essential Relativity*. Springer-Verlag, Heidelberg. 1977.

Riordan, Michael. *The Hunting of the Quark*. Simon & Schuster, New York. 1987.

Robb, Alfred Arthur. 'The Don of the Day' in *The American Physics Teacher*. Volume 7, Number 3. June 1939.

Robb, Alfred Arthur. 'The Revolution of the Corpuscle' in *The American Physics Teacher*. June 1939.

Robbins, Tom. *Even Cowgirls Get the Blues*. Houghton Mifflin Company, Boston. 1976.

Roberts, Michael. 'Notes on θ, ϕ, and ψ' in *The New Statesman*. March 23, 1935.

Roberts, Michael and Thomas, E.R. *Newton and the Origin of Colours*. G. Bell & Sons, London. 1931.

Roberts, Roylston. *Serendipity*. Wiley, New York. 1989.

Robinson, Arthur. 'Quantum Mechanics Passes Another Test' in *Science*. Volume 217, Number 4558. July 30, 1982.

Robinson, Edwin Arlington. *Collected Poems of Edwin Arlington Robinson*. The Macmillan Company, New York. 1952.

Robinson, Howard A. 'The Challenge of Industrial Physics' in *Physics Today*. June 1948.

Rogers, Eric M. *Physics for the Inquiring Mind*. Princeton University Press, Princeton. 1960.

Rogers, Eric M. *Astronomy for the Inquiring Mind*. Princeton University Press, Princeton. 1982.

Romanoff, Alexis L. *Encyclopedia of Thoughts*. Ithaca Heritage Books, Ithaca. 1957.

Rossetter, Jack C. 'Mathematical Notes' in *The Mathematics Teacher*. Volume 43. November 1950.

Rowan-Robinson, Michael. *Our Universe: An Armchair Guide*. W.H. Freeman, New York. 1990.

Royce, Josiah. *The World and the Individual*. Macmillan & Co., London. 1901.

Rucker, Rudy. *The Fourth Dimension*. Houghton Mifflin, Boston. 1984.

Ruderman, M.A. and Rosenfeld, A.H. 'An Explanatory Statement on Elementary Particle Physics' in *American Scientist*. Volume 48, Number 2. June 1960.

Ruffini, Remo and Wheeler, John. 'Relativistic Cosmology and Space Platforms' in *Proceedings of Conference on Space Physics*. ESRO, Paris. 1971.

Russell, Bertrand. *Introduction to Mathematical Philosophy*. George Allen and Unwin Ltd., London. 1920.

Russell, Bertrand. *Mysticism and Logic and Other Essays*. Longmans Green and Co., London. 1918.

Russell, Bertrand. *Religion and Science*. Oxford University Press, London. 1960.

Russell, Bertrand. *The ABC of Atoms*. Kegan Paul, Trench, Trubner & Co., Ltd., London. 1927.

Russell, Bertrand. *The ABC of Relativity*. George Allen & Unwin, Ltd., London. 1964.

Russell, Bertrand. *The Analysis of Matter*. Dover Publications, New York. 1954.

Russell, Bertrand. *The Principles of Mathematics*. G. Allen & Unwin, London. 1937.

Russell, Bertrand. *The Scientific Outlook*. W.W. Norton & Co., Inc., New York. 1931.

Russell, Bertrand. *Why I am not a Christian*. Allen & Unwin, New York. 1957.

Russell, H.N. 'Where Astronomers Go When They Die' in *Scientific American*. Volume 149. 1933.

Russell, Peter. *All for the Wolves*. Anvil Press Poetry, London. 1984.

Rutherford, Ernest. 'Atom Powered World Absurd, Scientists Told' in *NY Herald Tribune*. September 12, 1933.

Rutherford, Ernest. *Rutherford at Manchester*. W.A. Benjamin, New York. 1962.

Rutherford, Mark. *Last Pages from a Journal*. Oxford University Press, London. 1915.

Rutherford, Mark. *More Pages from a Journal*. Oxford University Press, London. 1910.

Sagan, Carl. *Cosmos*. Random House, New York. 1980.

Sagan, Carl. 'God and Carl Sagan: Is the Cosmos Big Enough for Them?' in *US Catholic*. May 1981.

Saint Augustine. *The Confessions* in *Great Books of the Western World*. Translated by Edward Bouberie Pusey. Volume 18. Encyclopædia Britannica, Inc., Chicago. 1952.

Sandage, A. *The Hubble Atlas of Galaxies*. Carnegie Institute of Washington, Washington D.C. 1961.

Santayana, George. *The Realm of Matter*. Constable and Company Ltd., London. 1930.

Sappho. *Sappho* edited by Mary Barnard. University of California Press, Berkeley. 1958.

Sayers, Dorothy L. *The Unpleasantness at the Bellona Club*. Harcourt, Brace and Company, New York. 1937.

Schid, Alfred. 'The Clock Paradox in Relativity Theory' in *American Mathematics Monthly 66*. January 1959.

Schlipp, Paul A. *Albert Einstein, Philosopher-Scientist*. Tudor Publishing Co., New York. 1951.

Schrödinger, Erwin. *Science Theory and Man*. George Allen and Unwin Ltd., London. 1935.

Schrödinger, Erwin. *Collected Papers on Wave Mechanics*. Blackie & Sons, London. 1929.

Schuster, Arthur. 'Potential Matter—A Holiday Dream' in *Nature*. Volume 58, Number 367. 1898.

Sciama, D.W. *The Unity of the Universe*. Doubleday & Company, Inc., Garden City. 1959.

Seaton, Ray and Hay, Will. *Good Morning Boys*. Barrie & Jenkins, London. 1978.

Seldes, George. *The Great Quotations*. A Caesar-Stuart Book, New York. 1960.

Selleri, Franco. *Quantum Mechanics Versus Local Realism*. Plenum Press, New York. 1988.

Seneca. *Hercules Furens*. Munroe, Boston. 1845.

Seneca. *Naturales Questiones*. Volume III. Harvard University Press, Cambridge. 1971.

Serviss, Garrett P. *Astronomy with the Naked Eye*. Harper & Brothers, New York. 1908.

Shakespeare, William. *Anthony and Cleopatra* in *Great Books of the Western World*. Volume 27. Encyclopædia Britannica, Inc., Chicago. 1952.

Shakespeare, William. *Hamlet, Prince of Denmark* in *Great Books of the Western World*. Volume 27. Encyclopædia Britannica, Inc., Chicago. 1952.

Shakespeare, William. *Julius Caesar* in *Great Books of the Western World*. Volume 27. Encyclopædia Britannica, Inc., Chicago. 1952.

Shakespeare, William. *Macbeth* in *Great Books of the Western World*. Volume 27. Encyclopædia Britannica, Inc., Chicago. 1952.

Shakespeare, William. *Othello, Moor of Venice* in *Great Books of the Western World*. Volume 27. Encyclopædia Britannica, Inc., Chicago. 1952.

Shakespeare, William. *Sonnets* in *Great Books of the Western World*. Volume 27. Encyclopædia Britannica, Inc., Chicago. 1952.

Shakespeare, William. *Troilus and Cressida* in *Great Books of the Western World*. Volume 27. Encyclopædia Britannica, Inc., Chicago. 1952.

Shapley, Harlow and Upton, Winslow. 'Harvard Observatory Pinafore' in *Popular Astronomy*. Volume 38, Number 3. March 1930.

Shaw, George Bernard. *In Good King Charles's Golden Days*. Constable & Co., London. 1939.

Sheehan, William. *Planets & Perception*. The University of Arizona Press, Tucson. 1988.

Shelley, Percy Bysshe. *The Complete Poetical Works of Shelley*. Houghton Mifflin Co., Boston. 1901.

Shirer, W.L. *The Rise and Fall of the Third Reich: A History of Nazi Germany*. Simon and Schuster, New York. 1960.

Shive, John N. and Weber, Robert L. *Similarities in Physics*. John Wiley & Sons, New York. 1982.

Silk, Joseph. *Cosmic Enigmas*. The AIP Press, Woodbury. 1994.

Silk, Joseph. *The Big Bang*. W.H. Freeman and Company, San Francisco. 1980.

Singer, Isaac Bashevis. *A Crown of Feathers*. Farrar, Straus and Giroux, New York. 1970.

Singh, Jagit. *Great Ideas of Modern Mathematics*. Dover Publications, New York. 1959.

Smart, Christopher. *Poems by Christopher Smart*. Princeton University Press, Princeton. 1950.

Smith, Alice and Weiner, Charles. *Robert Oppenheimer, Letters and Reflections*. Harvard University Press, Cambridge. 1980.

Smith, E.E. *Masters of the Vortex*. Pyramid, New York. 1968.

Smith, G. Gregory. *Elizabethan Critical Essays*. Volume II. Oxford University Press, London. 1904.

Smith, H.J.S. 'Presidential Address, British Association for the Advancement of Science' in *Nature*. Volume 8, 1873.

Smith, Logan. *Trivia*. Doubleday, Doran & Company, Inc., New York. 1917.

Smith, Robert W. *The Space Telescope*. Cambridge University Press, Cambridge. 1989.

Smith, Sydney. *The Wit and Wisdom of Sydney Smith*. G.P. Putnam's Sons, New York. No date.

Smoot, George. *Wrinkles in Time*. William Morrow and Company, Inc., New York. 1993.

Snow, C.P. *The Two Cultures and The Scientific Revolution*. Cambridge University Press, New York. 1959.

Sommerfeld, Arnold. *Wave-Mechanic*. Methuen & Co., Ltd., London. 1930.

Spencer-Brown, George. *Laws of Form*. Allen & Unwin, London. 1969.

Spenser, Edmund. *The Complete Poetical Works of Edmund Spenser*. Houghton Mifflin Co., New York. 1908.

Stallo, John Bernard. *The Concepts and Theories of Modern Physics*. 4th Edition. London. 1900.

Standen, Anthony. *Science is a Sacred Cow*. E.P. Dutton and Company, Inc., New York. 1950.

Stapledon, Olaf. *Last and First Men & Star Maker*. Dover Publications, Inc., New York. 1968.

Stedman, Edmund Clarence. *The Poems of Edmund Clarence Stedman*. Houghton Mifflin Co., Boston. 1908.

Stenger, Victor J. *Not By Design*. Prometheus Books, Buffalo. 1988.

Stern, Lawrence. *Tristram Shandy*. The Modern Library, New York. 1928.

Stevenson, Robert Louis. 'Fables' in *The Works of Robert Louis Stevenson*. Charles Scribner's Sons, New York. 1921.

Stewart, Dugald. *Elements of the Philosophy of the Human Mind*. Garland Publications, New York. 1971.

Stewart, Ian. *Does God Play Dice?* Basil Blackwell Inc., Cambridge. 1990.

Stone, Samuel John. 'Soliloquy of a Rationalistic Chicken' in *Harper's Monthly*. September 1875.

Stone, Wilfred. *Religion and Art of William Hale White*. AMS Press, New York. 1967.

Strutt, Robert John. *Life of John William Strutt: Third Baron Rayleigh*. The University of Wisconsin Press, Madison. 1968.

Stuart, Edward D., Cal-Or, Benjamin and Brainard, Alan J. *A Critical Review of Thermodynamics*. Mono Book Corp., Baltimore. 1970.

Sullivan, J.W.N. *The Bases of Modern Science*. Doubleday, Doran & Co., Inc., Garden City. 1929.

Sullivan, W.T, III. *Classics in Radio Astronomy*. Reidel Publishing Co., Hingham. 1982.

Swedenborg, Emanuel. *Concerning the Earths in Our Solar System, Which are Called Planets, and Concerning the Earths in the Starry Heavens*. Printed and sold by R. Hindmarsh, London. 1787.

Swift, Jonathan. *Gulliver's Travels* in *Great Books of the Western World*. Volume 36. Encyclopædia Britannica, Inc., Chicago. 1952.

Swift, Jonathan. *Satires and Personal Writings*. Oxford University Press, New York. 1932.

Swift, Jonathan. *The Portable Swift*. Viking Press, New York. 1948.

Sylvester, J.J. *On a Theorem Connected with Newton's Rule*. University Press, Cambridge. 1909.

Synge, J.L. *Relativity: The Special Theory*. North-Holland Publishing Co., Amsterdam. 1956.

Szent-Gyorgyi, Albert. 'Teaching and the Expanding Knowledge' in *Science*. Volume 146, Number 3649. December 4, 1964.

Szilard, Leo. *Leo Szilard: His Version of the Facts*. MIT Press, Cambridge. 1978.

Tagore, Rabindranath. *Our Universe*. Jaico Publishing House, Bombay. 1969.

Taube, M. *Evolution of Matter and Energy*. Springer-Verlag, New York. 1985.

Taylor, Anne. *Rhymes for the Nursery*. Peter B. Gleason & Co., Hartford. 1813.

Taylor, John G. *The New Physics*. Basic Books, Inc., New York. 1972.

Teller, Edward. *Conversations on the Dark Secrets of Physics*. Plenum Press, New York. 1991.

Tem, Steve Marcus. 'The Physicist's purpose' in *The Umbral Anthology of Science Fiction Poetry*. Edited by Steve Rasnic Tem. Umbral Press, Denver. 1982.

Tennyson, Alfred. *Poems of Tennyson*. Oxford University Press, London. 1912.

Tennyson, Alfred. *Alfred Tennyson's Poetical Works*. Oxford University Press, London. 1953.

Thiel, R. *And There was Light*. Alfred A. Knopf, New York. 1957.

Thompson, Silvanus P. *The Life of William Thomson, Baron Kelvin of Largs*. Volume 2. Macmillan, London. 1910.

Thompson, Silvanus P. *Calculus Made Easy*. The Macmillan Company, New York. 1929.

Thomsen, Dietrick E. and Eberhart, Jonathan. 'The Pulsar's Pindar' in *Science News*. Volume 93. 15 June 1968.

Thomson, James. *The Complete Poetical Works of James Thomson*. Oxford University Press, London. 1908.

Thomson, J.J. 'Eclipse Showed Gravity Variation: Hailed as Epochmaking' in *The New York Times*. November 9, 1919.

Thomson, J.J. *The Corpuscular Theory of Matter*. Charles Scribner's Sons, New York. 1907.

Thomson, William. 'On the Size of Atoms' in *Nature*. Volume 1. 1870.

Thomson, William. *Popular Lectures and Addresses*. Macmillan and Co., London. 1891.

Thoreau, Henry D. *Journal*. Volume 3: 1848–1851. Princeton University Press, Princeton. 1990.

Thoreau, Henry D. *The Journal of Henry D. Thoreau*. Edited by Bradford Torrey and Francis H. Allen. Dover Publications, Inc., New York. 1962.

Thoreau, Henry David. *Walden*. Published for the Classics Club by W.J. Black, New York. 1942.

Thorne, Kip S. 'The Search for Black Holes' in *Cosmology + 1*. W.H. Freeman and Company, San Francisco. 1977.

Thurber, James. *Many Moons*. Harcourt, Brace and Company, New York. 1943.

Tolstoy, Leo. *War and Peace* in *Great Books of the Western World*. Volume 51. Encyclopædia Britannica, Inc., Chicago. 1952.

Toulmin, S. and Goodgield, J. *The Fabric of the Heavens*. Harper & Row, New York. 1961.

Trimble, V. 'Dark Matter in the Universe: Where, What and Why?' in *Contemporary Physics*. Volume 29. 1988.

Truesdell, Clifford A. *Six Lectures on Modern Natural Philosophy*. Springer-Verlag, Berlin. 1966.

Tsu, Chuang. *Inner Chapters*. Translated by Gia-Fu Feng and Jane English. Alfred A. Knopf, New York. 1974.

Tukey, John W. 'The Future of Data Analysis' in *The Annals of Mathematical Statistics*. Volume 33, Number 1. March 1962.

Turner, H.H. 'Halley's Comet' in *The Mathematical Gazette*. Volume VI, Number 91. March 1911.

Twain, Mark. *Huckleberry Finn*. Harper, New York. 1931.

Twain, Mark. *Life on the Mississippi*. Harper & Brothers, London. 1906.

Updike, John. *Facing Nature*. Alfred A. Knopf, New York. 1985.

Updike, John. *Roger's Version*. Alfred A. Knopf, New York. 1986.

Updike, John. *The Centaur*. Alfred A. Knopf, New York. 1990.

Updike, John. *Tossing and Turning*. Alfred A. Knopf, New York. 1977.

Updike, John. *Telephone Poles and Other Poems*. Alfred A. Knopf, New York. 1964.

Uspenskii, Petr Demianovich. *Tertium Organum*. Vintage Books, New York. 1970.

Vaihinger, H. *The Philosophy of 'As if'*. Harcourt, Brace & Company, Inc., New York. 1925.

van der Waerden. B.L. *Sources of Quantum Mechanics*. North-Holland Publishing Company, Amsterdam. 1967.

van Loon, Hendrik. *The Story of Mankind*. Garden City Publishing Company, Inc., Garden City. 1926.

Van Over, Raymond. *Sun Songs*. New American Library, New York. 1980.

Van Sant, Gus. Screenplay of *Even Cowgirls Get the Blues*. Faber & Faber, London. 1993.

Veblen, Oswald. 'Geometry and Physics' in *Science*. February 2, 1923.

Virgil. *The Aeneid* in *Great Books of the Western World*. Translated by James Rhoades. Volume 13. Encyclopædia Britannica, Inc., Chicago. 1952.

Voltaire. *Candide*. Boni and Liveright, New York. 1918.

Voltaire. *The Best Known Works of Voltaire*. Blue Ribbon Books, New York. 1940.

von Mises, Richard. *Probability, Statistics, and Truth*. Dover Publications, New York. 1957.

von Weisacker, C.F. *The World View of Physics*. Translated by Majorie Grene. Routledge and Kegan Paul, Ltd., London. 1952.

Vyasa. *The Mahabharata of Vyasa*. Translated by P. Lal. Vikas Publishing House, New Delhi. 1980.

Walcott, Derek. *The Star Apple Kingdom*. Jonathan Cape, London. 1979.

Walters, Marcia C. *The Physics Teacher*. Volume 5, Number 8. November 1967.

Warren, Henry White. *Recreations in Astronomy: with directions for practical experiments and telescopic work*. Chautauqua Press, New York. 1879.

Weaver, Jefferson Hane. *The World of Physics*. Volumes I, II and III. Simon and Schuster, New York. 1987.

Weber, Renée. *Dialogues with Scientists and Sages: The Search for Unity*. Routledge and Kegan Paul, London. 1986.

Weber, Robert L. *Science with a Smile*. Institute of Physics Publishing, Bristol. 1992.

Weinberg, Steven. *The First Three Minutes*. Basic Books, Inc., New York. 1977.

Wells, H.G. *28 Science Fiction Stories of H.G. Wells*. Dover Publications, Inc. 1952.

Wells, H.G. *Experiment in Autobiography*. The Macmillan Company, New York. 1934.

Weyl, Hermann. *Mind and Nature*. University of Pennsylvania Press, Philadelphia. 1934.

Weyl, Hermann. *Space, Time, and Matter*. Translated by Henry L. Brose. Dover Publications, Inc., New York.

Weyl, Hermann. *Symmetry*. Princeton University Press, Princeton. 1952.

Weyl, Hermann. *The Theory of Groups and Quantum Mechanics*. Methuen & Co., Ltd., London. 1931.

Wheeler, J.A. *A Journey into Gravity and Spacetime*. Freeman, New York. 1990.

Wheeler, J.A. 'Geometrodynamics and the Issue of the Final State' in *Relativity, Groups, and Topology*. Edited by C. deWitt and B. DeWitt. Gordon and Breach, New York. 1964.

Wheeler, J.A. 'Our Universe: The Known and the Unknown' in *American Scientist*. Volume 56.

Wheeler, J.A. 'The Quantum and the Universe' in *Relativity, Quanta and Cosmology*. Edited by M. Pantaleo and F. de Finis. Volume II. Johnson Reprint Corporation, New York. 1979.

Wheeler, J.A. 'The Computer and the Universe' in *International Journal of Theoretical Physics*. Volume 21, Numbers 6/7. 1982.

Wheeler, J.A. 'Beyond the Black Hole' in *Some Strangeness in Proportion*. Edited by Harry Wolf. Addison-Wesley, Reading. 1980.

Wheeler, J.A., Rees, M. and Ruffini, R. *Black Holes, Gravitational Waves, and Cosmology*. Gordon and Breach, New York. 1974.

Wheeler, J.A., Thorne, K.S. and Misner, C. *Gravitation*. W.H. Freeman & Co., Salt Lake City. 1973.

Whewell, W. *Astronomy and General Physics*. W. Pickering, London. 1833.

Whewell, W. *History of the Inductive Sciences*. Volume II. D. Appleton, New York. 1890.

Whewell, W. *The Philosophy of the Inductive Sciences*. Volume I. John W. Parker, London. 1847.

White, Kirk. *Works of Gray, Blair, Beattie, Collins, Thomson and Kirke White*. James Blackwood & Co., London. 1883.

White, Stephen. 'A Newsman Looks at Physicists' in *Physics Today*. May 1948.

Whitehead, Alfred North. *Adventures of Ideas*. Cambridge University Press, Cambridge. 1933.

Whitehead, Alfred North. *An Introduction to Mathematics*. Oxford University Press, London. 1948.

Whitehead, Alfred North. *Modes of Thought*. The Macmillan Company, New York. 1938.

Whitehead, Alfred North. *Science and the Modern World*. The Macmillan Company, New York. 1929.

Whitehead, Alfred North. *The Aims of Education*. The Macmillan Company, New York. 1929.

Whitehead, Alfred North. *The Concept of Nature*. Cambridge University Press, Oxford. 1926.

Whitehead, Alfred North. *The Principles of Relativity*. Cambridge University Press. 1922.

Whitehead, Alfred North. 'The Idealistic Interpretations of Einstein's Theory' in *Proceedings of the Aristotelian Society*. N.S. Volume 22.

Whitman, Walt. *Complete Poetry and Collected* Prose. The Library of America, New York. 1982.

Whyte, Lancelot Law. *Essays on Atomism: from Democritus to 1960*. Harper & Row, New York. 1961.

Wickstrom, Lois. *Pandora*. Sproing, Inc. 1979.

Wilczek, Frank and Devine, Betsy. *Longing for the Harmonies*. W.W. Norton & Company, New York. 1988.

Wiener, Norbert. *God and Golem, Inc*. The M.I.T. Press, Cambridge. 1964.

Wigner, Eugene. *Symmetries and Reflections*. Greenwood Press, Westport. 1967.

Wilde, Oscar. *The Importance of Being Earnest*. Appeal to Reason, Girard. 1921.

Wilde, Oscar. *Lady Windermere's Fan*. John Roberts Press, London. 1973.

Wilkins, John. *The Discovery of a World in the Moone*. Printed by E.G. for Michael Sparke and Edward Forrest. London. 1638.

Williams, Sarah. *The Best Loved Poems of the American People*. Garden City Books, Garden City. 1936.

Williams, Tennessee. *Cat on a Hot Tin Roof*. A New Directions Book, New York. 1975.

Wilmot, John. *Collected Works of John Wilmot Earl of Rochester*. The Nonesuch Press, WC. 1926.

Wilson, David Alec. *Carlyle on Cromwell and Others*. Kegan Paul, Trench, Trubner & Co., Ltd., London. 1925.

Wilson, D.A. and MacArthur, D.W. *Carlyle in Old Age (1865–1881)*. Kegan Paul, Trench, Trubner & Co., Ltd., London. 1934.

Winsor, Frederick. *The Space Child's Mother Goose*. Simon and Schuster, New York. 1958.

Wittgenstein, Ludwig. *Tractatus Logico-Philosophicus*. Routledge & Kegan Paul, London. 1933.

Wolf, Alan. *Parallel Universes*. Simon & Schuster, New York. 1988.

Wolf, Fred Alan. *Star Wave*. The Macmillan Company, New York. 1984.

Wolkomir, Richard. 'Quark City' in *Omni*. February 1984.

Wordsworth, William. *The Poetical Works of William Wordsworth*. Oxford University Press, London. 1913.

Wright, Edward. *The Description and Use of the Sphaere*. Theatrum Orbis Terrarum Ltd., Amsterdam. 1969.

Yang, C.N. *Elementary Particles: A Short History of some Discoveries in Atomic Physics*. Princeton University Press, Princeton. 1961.

Yeats, William Butler. *The Collected Poems of W.B. Yeats*. The Macmillan Company, New York. 1935.

Young, Edward. *Night Thoughts*. Edited by Stephen Cornfford. Cambridge University Press, Cambridge. 1989.

Young, Joshua. 'Physics Poems' in *The Physics Teacher*. Volume 20, Number 9. December 1982.

Zee, A. *Fearful Symmetry*. The Macmillan Company, New York. 1986.

Zee, A. *An Old Man's Toy*. The Macmillan Company, New York. 1989.

Zeilik, Michael. *Astronomy: The Evolving Universe*. Harper & Row, New York. 1982.

Zimmer, H. *Myths and Symbols in Indian Art and Civilization*. Pantheon Books, New York. 1946.

Zolynas, Al. *The New Physics*. Wesleyan University Press, Middletown. 1979.

PERMISSIONS

Grateful acknowledgment is made to the following for their kind permission to reprint copyright material. Every effort has been made to trace copyright ownership but if, inadvertently, any mistake or omission has occurred, full apologies are herewith tendered.

Full reference to authors and the titles of their works are given under the appropriate quotations.

A JOURNEY INTO GRAVITY AND SPACETIME by John Archibald Wheeler. Reprinted by permission of the publisher, W.H. Freeman and Company, New York.

CONJECTURES AND REFUTATIONS: THE GROWTH OF SCIENTIFIC KNOWLEDGE by Karl Popper. Reprinted by permission of the publisher, HarperCollins Publishers, London.

COSMOLOGY: SCIENCE OF THE UNIVERSE by Edward Harrison. Reprinted by permission of the publisher, Cambridge University Press.

CRYSTALS AND X-RAYS by Dame Kathleen Lonsdale. Reprinted by permission of the publisher, Prentice-Hall, Inc., Englewood Cliffs.

ESSENTIAL RELATIVITY by Wolfgang Rindler. Reprinted by permission of the publisher, Springer-Verlag, New York.

FRONTIERS OF ASTRONOMY by Fred Hoyle. Reprinted by permission of the Estate of Fred Hoyle.

INTRODUCTION TO ASTRONOMY by Cecilia Payne-Gaposchkin. Reprinted by permission of the publisher, Prentice-Hall, Inc., Englewood Cliffs.

MATHEMATICS IN WESTERN CULTURE by Morris Kline. Reprinted by permission of the publisher, Oxford University Press, New York.

MATTER, EARTH AND SKY by George Gamow. Reprinted by permission of the publisher, Prentice-Hall, Inc., Englewood Cliffs.

NIGHT THOUGHTS by Edward Young. Reprinted by permission of the publisher, Cambridge University Press.

OCTOBER THE FIRST IS TOO LATE by Fred Hoyle. Reprinted by permission of the publisher, HarperCollins Publishers, New York.

OUT ON A LIMERICK by Bennett Cerf. Reprinted by permission of Phyllis Cerf Wagner.

PHYSICS AND BEYOND by Werner Heisenberg. Reprinted by permission of the publisher, HarperCollins Publishers, New York.

PHYSICS AND PHILOSOPHY by Werner Heisenberg. Reprinted by permission of the publisher, HarperCollins Publishers, New York.

SIDE EFFECTS by Woody Allen. Reprinted by permission of the publisher, Random House, Inc., New York.

SUPERSTRINGS by P.C.W. Davies and J. Brown. Reprinted by permission of the publisher, Cambridge University Press.

THE BIG BANG by Joseph Silk. Reprinted by permission of the publisher, W.H. Freeman and Company, New York.

THE EMPEROR'S NEW MIND by Roger Penrose. Reprinted by permission of the publisher, Oxford University Press, New York.

THE FABRIC OF THE HEAVENS by Stephen Toulmin and June Goodfield. Reprinted by permission of the publisher, HarperCollins Publishers, New York.

THE FEYNMAN LECTURES ON PHYSICS VOL. 1 by R. Feynman, R. Leighton and M. Sands. Reprinted by permission of the publisher, Addison-Wesley Longman, Inc., Reading.

THE FEYNMAN LECTURES ON PHYSICS VOL. 2 by R. Feynman, R. Leighton and M. Sands. Reprinted by permission of the publisher, Addison-Wesley Longman, Inc., Reading.

THE FEYNMAN LECTURES ON PHYSICS VOL. 3 by R. Feynman, R. Leighton and M. Sands. Reprinted by permission of the publisher, Addison-Wesley Longman, Inc., Reading.

THE FIRST THREE MINUTES by Steven Weinberg. Reprinted by permission of the publisher, HarperCollins Publishers, New York.

THE LADY'S NOT FOR BURNING by Christopher Fry. Reprinted by permission of the publisher, Oxford University Press, New York.

THE LAST THREE MINUTES by Paul Davies. Reprinted by permission of the publisher, HarperCollins Publishers, New York.

THE NATURE OF THE UNIVERSE by Fred Hoyle. Reprinted by permission of the Estate of Fred Hoyle.

THE PHILOSOPHICAL IMPACT OF CONTEMPORARY PHYSICS by Milič Čapek. Reprinted by permission of the publisher, Van Nostrand Reinhold, New York.

THE PHILOSOPHY OF PHYSICAL SCIENCE by A.S. Eddington. Reprinted by permission of the publisher, Cambridge University Press.

THE PRIMEVAL ATOM by G. Lemaître. Reprinted by permission of the publisher, Prentice-Hall, Inc., Englewood Cliffs.

SUBJECT BY AUTHOR INDEX

Longair, Malcolm
　Just keep away from the black
　　hole garbage can..., 35
　(Cartoon, 298)
Ruffini, Remo
　A black hole has no hair, 36
Thorne, Kip S.
　Of all the conceptions of the
　　human mind...perhaps the
　　most fantastic is the black
　　hole..., 36
Unknown
　A black hole is where God
　　divides by zero, 36
Wheeler, John A.
　...go down the black hole..., 36
　...pondering the paradox of
　　gravitational collapse..., 36
Boötes
Manilius, Marcus
　And next Boötes comes..., 52

-C-

calculation
Buck, Pearl S.
　If, after calculation...it were
　　proved..., 37
Dirac, Paul Adrien Maurice
　...I understand an equation
　　when I can predict the
　　properties of its solutions...,
　　37
Einstein, Albert
　If A is success in life..., 38
Graham, L.A.
　But figure out β, 38
Lillich, Robert
　Divergence B, it's plain to see, is
　　zero..., 150
Wittgenstein, Ludwig
　Calculation is not an experiment,
　　38
calculations
Bloch, Felix

Erwin with his psi can
　do/Calculations quite a few,
　37
Einstein, Albert
　Your calculations are correct, but
　　your physics is abominable,
　　38
Canis Major
Aratus
　...In his fell haw/Flames a star
　　above all others..., 278
Frost, Robert
　The great Overdog..., 50
cause and effect
Einstein, Albert
　...except when observable facts
　　ultimately appear as causes
　　and effects, 39
Heisenberg, Werner
　The chain of cause and effect...,
　　39
Russell, Bertrand
　Thus geometry and causation
　　become inextricably
　　intertwined, 39
chaos
Adams, Henry
　Chaos was the law of nature...,
　　40
Aurelius, Marcus [Antoninus]
　Either it is a well arranged
　　universe or a chaos huddled
　　together..., 40
Dylan, Bob
　Chaos is a friend of mine, 40
Ford, Joseph
　...vast wasteland of chaos..., 40
Frost, Robert
　Let chaos storm!, 41
Harrison, Edward Robert
　...in the ultimate and
　　unimaginable chaos..., 41
Miller, Henry
　A Chaos whose order is beyond
　　comprehension, 41

-I-

infinite

-M-

Now Mars dominates the
Heavens, 232
Tennyson, Alfred Lord
Mars as he glowed like a ruddy
shield on the Lion's breast,
234
mathematical
Carlyle, Thomas
It is a mathematical fact..., 147
Montague, W.P.
...a crystalline participate
of purely mathematical
relations, 150
Unknown
I'm very well acquainted with
matters mathematical..., 151
mathematical construction
Einstein, Albert
Experience remains...the sole
criterion of the physical
utility of a mathematical
construction, 148
mathematical physicist
D'Abro, A.
Success has attended the efforts
of mathematical physicists...,
147
Petroski, Henry
Embedded in a matrix of
mistakes..., 151
mathematician
Whewell, William
...great mathematicians who
have laboured with such
wonderful success..., 151
mathematics
Bacon, Francis
As Physic advances...it will
require fresh assistance from
Mathematic, 147
Courant, Richard
Recent trends...weakened
the connection between
mathematics and physics...,
147

Dirac, Paul Adrien Maurice
...physics requires...a
mathematics that gets
continually more advanced,
148
...the beauty of a mathematical
theory of physics..., 148
...there will be no physical
treatise which is not
primarily mathematical, 148
Einstein, Albert
...as far as the propositions of
mathematics refer to reality,
they are not certain..., 256
...nature is the realization of
the simplest conceivable
mathematical ideas, 148
Don't worry about your
difficulties in mathematics...,
149
How can it be that
mathematics..., 149
Mathematics is only a means for
expressing the laws..., 149
One reason why mathematics
enjoys special esteem..., 149
Farrar, John
...in mathematical science...man
see things precisely as
God..., 149
Feynman, Richard P.
What is mathematics doing in
physics lectures?, 149
Gibbs, J. Willard
...the oldest of the scientific
applications of
mathematics..., 150
Hutten, Ernest H.
There are...no mathematical
models in physics..., 162
Jeans, Sir James Hopwood
...the pictures which science
draws of nature...are
mathematical pictures,
150

AUTHOR BY SUBJECT INDEX

matter, 154
space, 269
time, 323
Kasner, Edward (1878–1955)
Mathematician
infinity, 124
Keane, Bill (1922–?)
US cartoonist
heat, 112
Kepler, Johannes (1571–1630)
German astronomer
astronomy, 13
God, 104
universe, infinite, 341
Keyser, C.J. (1862–1947)
universe, 341
King, Alexander (1900–1965)
Newton, 108
Kirk, Captain James T. [Captain
USS Enterprise, Star Trek]
space, 270
Kirkup, James (1918–?)
Poet
Ursa Major, 51
Kirshner, Robert (–)
astronomer, 6
Kitaigorodski, Aleksander
Isaakovich (1914–?)
Russian physicist
theory, 312
Kitchiner, William (1775–1827)
telescopes, 303
Kline, Morris (1908–?)
US mathematician
matter, 154
Koestler, A. (1905–1983)
Hungarian-born British author
data, 62
law, 134
Kramers, Hendrick Anthony
(1894–1952)
Dutch physicist
quantum theory, 247
Kraus, Karl (–)
comet, 45

Krutch, Joseph Wood (1893–1970)
US writer
physicists, 200
Kudlicki, Andrzej (–)
red shift, 257
Kuhn, Thomas S. (1922–?)
US historian of science
physicist, 200
Kunitz, Stanley (1905–?)
universe, 341
Kunz, F.L. (–)
universe, 341

-L-
Ladenburg, Rudolf (–)
physicists, 200
Lamarck, Jean-Baptiste
(1744–1829)
French naturalist
atoms, 24
Lamb, Charles (1775–1834)
English poet
space, 270
Lambert, Heinrich (1728–1777)
Swiss-German mathematician
Milky Way, 159
Laplace, Pierre Simon (1749–1827)
French mathematician
black hole, 35
gravitational fluid, 108
hypothesis, 117
law, 134
Larrabee, Eric (–)
astronomy, 14
physics, 218
Latham, Peter M. (1789–1875)
fact, 88
Lawrence, D.H. (1885–1930)
English writer
relativity, 259
Lawson, Alfred William (–)
truth, 328
Leacock, Stephen (1869–1944)
Canadian humorist
astronomy, 14

-M-

Mach, Ernst (1838–1916)
Austrian physicist
  atoms, 26
  physics, 218
  relativity, 259
  theory, 313
Mackay, Charles (1814–1889)
  symmetry, 296
Maimonides, Moses (1135–1204)
Jewish scholastic philosopher
  physics, 218
  theory, 313
Mamula, Karl C. (–)
  force, 93
Manilius, Marcus (early 1st
    century AD)
Roman didactic poet
  Aries, 51
Mann, Thomas (1875–1955)
German author
  sun, 292
Mao Tse-tung (1893–1976)
Chinese Marxist theorist
  symmetry, 296
March, Robert H. (1937–?)
  physicists, 200
Marcus, Adrianne (–)
  physicist, 200
Marduk, Babylonian Sun-god
  constellation, 52
Marlowe, Christopher (1564–1593)
English dramatist
  knowledge, 131
Marquis, Don (1878–1937)
Writer
  Einstein, Albert, 68
  universe, 342
  vacuum, 357
Marr, Ray (–)
  hypothesis, 116
  knowledge, 130
Martin, Florence Holcomb (–)
  meteor, 158
Marton, Ladislaus (–)

optics, 183
Maxwell, James Clerk (1831–1879)
British physicist
  atoms, 25
  ether, 80
  experiment, 83
  force, 93
  light, 141
  physics, 218
  space, 270
  space–time, 272
  statistical, 289
Mayer, Julius Robert von
    (1814–1878)
German physicist
  heat, 112
Mayer, Maria Goeppert
    (1906–1972)
German physicist
  physics, 219
Mayer, Robert (1814–1878)
German physicist
  sun, 292
Mayes, Harlan Jr. (–)
  theory, 313
McAleer, Neil (1942–?)
  galaxy, 97
  universe, 342
McCrea, William Hunter (–)
Astronomer
  cosmos, 56
Meixner, J. (–)
  thermodynamics, 318
Mellor, J.W. (1873–1938)
  data, 62
Melneckuk, Theodore (–)
  quark, 250
Melville, Herman (1819–1891)
US author
  atom, 26
Mencken, H.L. (1880–1956)
US journalist
  astronomers, 6
  God, 104
  physics, 219

astronomer, 9
atom, 30
stars, 288
Wright, Edward (1587–1615)
planet, 235

-Y-

Yang, Chen N. (1922–?)
Chinese-born US theoretical physicist
quantum mechanics, 249
symmetry, 297
Yeats, William Butler (1865–1939)
Irish poet
stars, 288
time, 327
Young, Edward (1683–1765)
English poet
astronomer, 9
comet, 48
constellation, 54
matter, 156
particles, 190
stars, 288

Young, Joshua (–)
light, 143
motion, 168

-Z-

Zebrowski, George (–)
universe, 349
Zee, Anthony (–)
light, 143
mathematics, 152
physicists, 206
symmetry, 298
Zel'Dovich, Ya.B. (1944–?)
Russian cosmologist
astronomers, 55
asymmetry, 105
big bang, 33
Zolynas, Al (1945–?)
Poet
physics, 228
Zuni Creation Myth
universe, cosmogenesis, 354

Printed and bound by CPI Group (UK) Ltd, Croydon, CR0 4YY

17/10/2024

01775686-0019